Dissertations
in
American Economic History

This is a volume in the Arno Press collection

Dissertations in American Economic History

Advisory Editor
Stuart Bruchey

Research Associate
Eleanor Bruchey

See last pages of this volume for a complete list of titles.

THE YEOMAN FARMER AND WESTWARD EXPANSION OF U.S. COTTON PRODUCTION

James D. Foust

ARNO PRESS
A New York Times Company
New York – 1975

First publication in book form, Arno Press, 1975

DISSERTATIONS IN AMERICAN ECONOMIC HISTORY
ISBN for complete set: 0-405-07252-X
See last pages of this volume for titles.

Manufactured in the United States of America

Library of Congress Cataloging in Publication Data

Foust, James D
The yeoman farmer and westward expansion of U. S. cotton production.

(Dissertations in American economic history)
Originally presented as the author's thesis, University of North Carolina, 1968.
Bibliography: p.
1. Cotton growing--Southwestern states--History. 2. Farmers--Southwestern states--History. 3. Slavery in the United States--Extension to the Territories. I. Title. II. Series.
HD9075.3.F68 1975 338.1'7'3510975 75-2581
ISBN 0-405-07201-5

THE YEOMAN FARMER AND WESTWARD EXPANSION OF U.S. COTTON PRODUCTION

by

James D. Foust

A dissertation submitted to the faculty of the University of North Carolina at Chapel Hill in partial fulfillment of the requirements for the degree of Doctor of Philosophy in the Department of Economics

Chapel Hill

1968

TABLE OF CONTENTS

ACKNOWLEDGMENTS

In the preparation of this manuscript, my debts to others are so great and so varied that I am unable to acknowledge all of them.

At the University of North Carolina, Professor Robert E. Gallman has been both an inspiration and a patient taskmaster. Professors James H. Blackman and Thomas J. Orsagh have offered numerous suggestions which substantially improved major portions of this study.

Professor Ross M. Robertson of Indiana University gave considerable time, effort and encouragement to this project. I am indebted to the School of Business at Indiana for providing time and facilities to complete this research. The many secretaries at the School of Business who had a hand in typing numerous drafts were patient and helpful. William Young and Gary Lindsay provided the computer programming without which this manuscript could not have been written.

Additionally, this entire study is based upon a manuscript census sample selected by numerous researchers over several years financed by a National Science Foundation grant.

Finally, I am indebted to numerous economic historians who have taken the time to provide many helpful suggestions. Especially, the encouragement of Professor William N. Parker of Yale has been most inspiring.

Errors, omissions, and "fuzzy" thinking I claim for myself.

James D. Foust

LIST OF TABLES

LIST OF TABLES (Continued)

LIST OF TABLES (Continued)

LIST OF TABLES (Continued)

CHAPTER I

INTRODUCTION

In the writing of economic history, the role of the little man is often overlooked. Watt, Arkwright, Stephenson, Gregg and Rockefeller are all familiar figures; the John Does who make up the bulk of any significant historical movement remain unknown.[1] It is the purpose of this study to shed some light on the John Doe who took part in the westward expansion of the cotton economy in the southern United States. Although we shall not get to know him by name, we shall attempt to ascertain his role in this phase of the westward movement.

THE WESTWARD MOVEMENT

Writing in 1862, J. E. Cairnes characterized the slave society as consisting of three distinct classes--slave, slaveholders, and "an idle and lawless rabble."[2] Most historians have not accepted this position. Rather, they have "strongly maintained that the white people of the ante-bellum rural South were not divided into the simple categories of rich planter and poor white."[3] Despite their recognition that it is impossible to divide the white population into two very distinct classes, historians have attempted to evaluate the role of the small slaveholder and

---

[1]The late Professor Frank Owsley and his students at Vanderbilt have attempted to remedy this situation. See Frank L. Owsley, *Plain Folk of the Old South* (Chicago: Quadrangle Books, 1965).

[2]*The Slave Power: Its Character, Career and Probable Designs*, 2nd ed. (London: Macmillan and Co., 1863), p. 60.

[3]Owsley, *op. cit.*, p. 5.

non-slaveholder vis a vis the large slaveholder or plantation owner.[4] It is the purpose of the present study to evaluate the role of the non-slaveholding and small-slaveholding farmers in the westward expansion of the antebellum cotton economy.

Regarding the northern United States, the role of the yeoman farmer in westward expansion has been extolled. When picturing westward expansion into the Old Northwest, one sees sturdy pioneer families moving their few possessions over the mountains and down the rivers to cheap, fertile lands beyond. The wealthy farmer moving fine furnishings, extensive equipment and, perhaps, a labor force is absent.[5]

In the South, the picture is not so clear. In some descriptions, a sharp contrast between northern and southern expansion is drawn; southern expansion is characterized by a wave of large planters, slaves and household possessions in tow, sweeping into the rich cotton lands of Alabama, Mississippi and Louisiana.[6] Other writers characterize expansion as the movement of herders and small farmers into the new lands.

Professor Owsley takes the latter position and emphasizes the ease with which small farmers could acquire southern frontier land.

---

[4]See particularly Eugene D. Genovese, _The Political Economy of Slavery: Studies in the Economy & Society of the Slave South_ (New York: Pantheon Books, 1965); Lewis C. Gray, _History of Agriculture in the Southern United States to 1860_, 2 vols. (New York: Peter Smith, 1941); Owsley, _loc. cit._; Ulrich B. Phillips, _Life and Labor in the Old South_ (Boston: Little Brown and Company, 1963).

[5]See Frederick Jackson Turner, _The Frontier in American History_ (New York: Henry Holt and Company, 1920).

[6]This contrast is especially vivid in economic history textbooks. See Lester S. Levy and Roy J. Sampson, _American Economic Development_ (Boston: Allyn and Bacon, Inc., 1963), pp. 133-41 and Ross M. Robertson, _History of the American Economy_, 2nd ed. (New York: Harcourt Brace & World, Inc., 1964).

He suggests that

> . . . the opportunity of acquiring land was greater in the South than in the North. . . . At all times during the interval between the Revolution and the Civil War the combined Federal and state public domains in the South were greater than those open to settlement in the North, and the population was increasingly less.[7]

Ulrich B. Phillips takes a position which is somewhat similar to Professor Owsley agreeing that the small farmers in search of new opportunity were the first to follow trappers and herders into the frontier regions. However, according to Phillips, these small farmers were

> . . . followed in many areas by a tide of planters who sought new openings where their capital might be employed more advantageously than in the older areas where competition was more stringent . . .
>
> Some of the slaveholders moved away from the encroaching plantations and settled anew as yeoman farmers; others by thrift bought slaves and in time became planters; others simply held their own in spite of the disadvantage of competing with Negro labor in the same industry; while still others retrograded in the scale of life, drifted to the barren tracts, and lived from hand to mouth as anemic poor-whites. The planters, meanwhile, continued to encroach wherever they could upon the territory already occupied in part by the smaller industrial unit.[8]

Thus, Phillips pictures westward expansion as a two step process: first, the yeomen established themselves in a region; second, the large plantation owners moved in forcing the small farmers out. Professor Owsley disagrees with Phillips maintaining that the small farmers

---

[7]Owsley, op. cit., p. 17.

[8]Quoted in Fabian Linden, "Economic Democracy in the Slave South: An Appraisal of Some Recent Views," Journal of Negro History, XXXI (April, 1946). Reprinted in Harold D. Woodman (ed.), Slavery and the Southern Economy (New York: Harcourt, Brace & World, Inc., 1966), p. 148.

> . . . lived dispersed over all the arable regions of the South and were settled in considerable numbers on every type of soil adapted to agricultural uses except swamp and river lands. . . . The truth of the matter is that the plain farmers settled where they chose and stayed as long as it suited them.[9]

The picture of westward expansion in the southern United States is further clouded by Phillips' description of the various classes of people in different agricultural regions. When discussing "Planters of the Southwest" he describes eight planters, four of whom entered the region with either slaves, wealth or both. A fifth planter entered the region as a plantation manager after meeting bankruptcy as a ship-chandler. Although no mention is made of the other three planters bringing slaves or wealth to the region, they are not pictured as yeoman farmers who acquired wealth and slaves in the new lands.[10] In his discussion of "The Plain People" Phillips mentions twelve individuals or family units. Six of these "plain people" lived in Virginia, North Carolina, or Georgia; one each in Alabama, Tennessee and Missouri; two in Mississippi; and one unknown. Phillips devotes considerable discussion to the plain people of the Georgia piedmont, the "mountain coves" and the upland regions of Virginia and the Carolinas.[11]

The documents which Phillips chose to include in _Plantation and Frontier_ neglect the nonslaveholding farmer.[12] In Chapter X, "Migration," the migrations of six families are described. Only one family was

---

[9] Owsley, _op. cit._, p. 52.

[10] Phillips, _op. cit._, Chapter XIV.

[11] _Ibid._, Chapter XVII.

[12] _A Documentary History of American Industrial Society_, Vol. II: _Plantation and Frontier_ (Cleveland: The Arthur H. Clark Company, 1910).

slaveless. An additional document in the chapter laments ". . . the constant emigration to the great West of our most substantial citizens . . . ."[13] In a later chapter, a letter from a Doctor Hoxey in Cole's Settlement, Texas (1834) states: ". . . my immediate neighbors are all slave holders and cotton planters. . . ."[14] Finally, in a selection taken from Sir Charles Lyell's _A Second Visit to the United States_ (published in 1846) the author makes the observation that ". . . on board (a steamer bound from Mobile to New Orleans) were many 'movers', going to Texas with their slaves. . . ."[15] Thus, from the examples Phillips cites, one gets the impression that the wealthy, slave owning planters moved to the Southwest while the plain people remained in the older sections of the South especially in the uplands and mountain coves.

Two writers who have contributed much to the literature of the American Frontier and indirectly to westward migration are Frederich Jackson Turner and Fred A. Shannon.[16] Both Turner and Shannon describe the southern migration as primarily a movement of larger farmers and plantation owners. Shannon suggests that prior to 1860 ". . . the white yeoman farmers (in the West South Central region) were largely confined to subsistence crops on the less fertile acres. . . ."[17] Turner not only

---

[13] _Ibid._, pp. 196-7.

[14] _Ibid._, p. 253.

[15] _Ibid._, p. 255.

[16] Turner, _op. cit._ and _Frontier and Section_ (Englewood Cliffs: Prentice-Hall, Inc., 1961); Fred A. Shannon, _The Economic History of the United States_, Vol. V: _The Farmer's Last Frontier, Agriculture, 1860-1879_ (New York: Rinehart & Company, Inc., 1959).

[17] _Ibid._, pp. 98-9.

emphasizes the migration of larger farmers but contrasts the differences between North and South.

> What the New England States and New York were in the occupation of the Middle West, Virginia, the Carolinas, and Georgia were in the occupation of the Gulf States. But, as in the case of the Northwest, a modification of the original stock occurred in the new environment. A greater energy and initiative appeared in the new Southern lands; the pioneers' devotion to exploiting the territory in which he was placed transferred slavery from the patriarchal to the commercial basis.[18]

Turner's emphasis upon the importance of the slaveholding farmer in southwestward expansion is evidenced by the observations that ". . . along the Gulf Plains there developed the: "Cotton Kingdom", a Greater South with a radical program of slavery expansion mapped out by bold and aggressive leaders. . ." and that

> ". . . the Cotton Kingdom based on slavery found its center in the fertile soils along the Lower Mississippi and the black prairies of Georgia and Alabama, and was settled largely by planters from the old cotton lands of the Atlantic States. The Mississippi Valley had rejuvenated slavery, had given it an aggressive tone characteristic of Western life.[19]

Turner lends support to Phillips' assertion that the small farmers were displaced from the better lands suggesting that

> . . . with the spread of the cotton culture went the slave system and the great plantation. The small farmer in his log cabin, raising varied crops, was displaced by the planter raising cotton. . . . The unity of the back country was broken and the solid South was formed.[20]

Lewis Cecil Gray's monumental work on southern agriculture includes passages which support Owsley's position that small farmers could easily acquire land in the Southwest, Phillips' position that small farmers

---

[18] The Frontier in American History, p. 139. (emphasis by this writer)

[19] Ibid., pp. 174 and 198.

[20] Ibid., pp. 216-7.

largely remained in the mountain coves of the Southeast, and the Turner-Phillips' position that small farmers were displaced by the larger planters. Gray suggests that ". . . as a rule, pioneer planters entered a region in the wake of pioneer farmers but . . . there were instances of planters with from 50 to 100 slaves who established themselves in the midst of the forest."[21] Further, he cites the ". . . heavy drain (from Virginia) of yeomen farmers, farm laborers, and artisans who sought to escape from the uncomfortable environment of large plantations and the competitions of slave labor."[22] But Gray also mentions that ". . . after the Texas Revolution planters with their slaves began to move to Texas in large numbers, settling on the alluvial lands of the Brazos, Trinity, Colorado, and Red rivers."[23] And, "although the plantation system tended to emerge triumphant in new regions where conditions were favorable, there was a tendency toward a decline of the system in old planting regions, whence the planters and slaves were inevitably drawn to new regions."[24] Yet, Gray supports the observation that the smaller farmers were ". . . unable to resist the competitive power of slave labor organized under the plantation system (and were gradually) . . . compelled either to become great planters . . . or to reestablish a regime of rude self-sufficing economy in a region less favorable for commercial agriculture."[25]

---

[21]*Op. cit.*, p. 438.

[22]*Ibid.*, p. 914.

[23]*Ibid.*, p. 907.

[24]*Ibid.*, p. 445.

[25]*Ibid.*, p. 444.

The work of the above writers has likely contributed to the sharp contrast between northern and southern migration presented in the Levy and Sampson textbook cited earlier. Professor Douglas North's description of the antebellum southern economy also leaves one with the impression that westward expansion of cotton production was the work of wealthy, slaveowning planters. In describing westward expansion, Professor North suggests that ". . . a period of high cotton prices led to a surge of _planters_ _and_ _slaves_ into the new Southwest."[26]

Professor Ross M. Robertson is willing to attribute the initial movement into the Southwest to traders and trappers followed closely by yeoman farmers but adds that

> . . . almost simultaneously--and this is what clearly distinguishes the southern migration--came the planter, the man of substance, with his great household establishment and his slaves. The planters had sufficient means to acquire large acreages, and the small householders could compete in the best agricultural areas only with difficulty. . . . The pioneer and the farmer with a slave or two were continually tempted to sell out to the plantation owner . . . .[27]

Thus, although opinion is varied, many authors emphasize the fact that migration into the Southwest was characterized by the migration of substantial planters prior to, alongside, or immediately following the migration of yeoman farmers. These authors see the Southwest as a plantation economy in which the larger planter's dominance was nearly as great or, perhaps, even greater than it was in the Southeast. In short, they do not attribute to the southwestern frontier the same democratizing influence which is frequently attributed to the new lands in the Northwest.

---

[26] _Growth and Welfare in the American Past: A New Economic History_ (Englewood Cliffs: Prentice-Hall, Inc.: 1966), p. 93. (Emphasis supplied by this writer.)

[27] Robertson, _op. cit._, p. 111.

THE PRESENT STUDY

Relation of this Study to Other Work

The above brief description suffices to indicate the general lack of information regarding the small farmers' role in westward movement in the South. One reason for this lack of information is that ". . . the plain folk did not make records comparable to those of the planters. The letters they wrote were few, and their significant, explicit items fewer still."[28]

Professor Owsley and his students have attempted to provide us with more information about the plain folk and have discovered in the manuscript census returns a body of data in which the plain folk did leave a record comparable to that left by the large planters. In fact, so long as the census enumerator did a careful job, the record for a small farmer and a large plantation owner would contain exactly the same data. Fabian Linden, while praising the contribution of the Vanderbilt group's utilization of the census manuscripts, has criticized the statistical techniques which were employed.[29]

Recently, the opportunity to study the record left in the manuscript census by farmers, plain and not-so-plain, has been enhanced by the microfilming of the manuscript censuses of agriculture of fifteen Southern states for 1850, 1960, 1870, and 1880. These microfilms have been processed and collected by the University of North Carolina Library. The manuscript census of free and slave population are also on microfilm. (Copies of the population censuses may be obtained from National Archives.)

---

[28]Phillips, *op. cit.*, p. 340.

[29]Linden, *op. cit.*

By matching the data from each of the population schedules with the data from the agricultural schedules it is possible to reconstruct the individual farms as they appeared to the census enumerator.[30]

Professor Owsley's students, working in the late forties, noted that the matching procedure was ". . . slow and tedious, and the life of one person would scarcely be long enough to check and analyze the structure of land and slave ownership and the productions of agriculture in every county of a single large state through two censuses."[31] Consequently, they selected a sample of a very few counties (taken from five states) and considered mainly land and slave ownership. Today, the researcher has available additional microfilm plus high speed computers which can analyze the data rapidly. The matching procedure remains slow and tedious.

The recent availability of the manuscript census of agriculture and the opportunity to use computer techniques to analyze the tremendous quantity of data have resulted in a new assault on the manuscript census by Professors William N. Parker and Robert E. Gallman and their students at Yale and the University of North Carolina. This assault has yielded the matched sample of farms selected from the 1860 manuscript census described in Chapter II. The present work is both a portion of the Yale-University of North Carolina project and an extension of that work.

This study is an attempt to determine the participants in the westward migration into the Southwest during the antebellum period. Essentially, the study is a test of a hypothesis using a statistical

---

[30]The details of this matching process will be described in Chapter II below.

[31]Owsley, _op. cit._, p. 153.

sample. The hypothesis under consideration and the method used to evaluate its accuracy are described below.

The Hypothesis

Earlier in this chapter various descriptions of southwestern migration were presented. With the exception of Professor Owsley, the writers quoted characterize southwestward expansion as a migration of larger planters who either lead or *immediately* follow the migration of smaller farmers. This description, which seems to be the dominant one, is the basis for the hypothesis which is evaluated in this paper. This hypothesis can be stated in two parts:

(a) The migration into the new land in the Southwest was one of large planters leading, accompanying, or *immediately* following smaller farmers.

(b) In the case that the large planters immediately followed the smaller farmers, the smaller farmers were pushed out by larger planters who were able to buy the better land from them.

In the first part of the hypothesis the word "immediately" may need clarification. For the purposes of this study the word "immediately" and Robertson's phrase "almost simultaneously" are taken to mean: "without interval of time." Migration involves a flow of people over time. In the antebellum South this flow certainly took place for at least half a century although there was some ebb and flow to be sure. Because of the length of time involved in settling any region, some of the "leading" group in the migration are certain to be accompanies by members of the "following" group, especially if the time lag between the two groups is quite small. Thus, for our purposes it seems reasonable to assume that a migration characterized by the movement of a "leading" group followed *immediately* by the "lagging" group is indistinguishable from a migration in which the two groups move simultaneously. Thus, in order to test this part of the hypothesis we might ask the question: "Did the small farmers lead the way into the newer regions?" If our answer to this question is "yes", we can reject the hypothesis.

Testing the Hypothesis

If the first part of the hypothesis were true, i.e., if large planters led the migration, one would expect to find relatively more large planters in the Southwest than in the Southeast. Further, one would expect that the fraction of land and other wealth held by larger planters in the Southwest would exceed the fraction held by the same group in the Southeast. Finally, one would expect the fraction of southwestern crops produced by large planters to exceed their share of southeastern crop production. Or, conversely, if the hypothesis is correct, one would expect that the small farmer would produce a smaller fraction of the total output and hold a smaller percentage of total wealth in the Southwest than in the Southeast. And, one would expect this distribution to persist for some period of time. That is, it seems highly unlikely that one would find a situation where large planters led the migration into the region but the region itself contained mostly small farmers or small and large farmers in the same proportions as they existed in regions from whence they emigrated. In fact, if two mutually exclusive but exhaustive regions are compared, the resulting differences are magnified because any increase in the one becomes a decrease in the other.

The first part of the hypothesis will be tested by examining the distribution of farm sizes and slaveholdings in the newer (Southwestern) and older (Southeastern) regions in the census years 1850 and 1860. If the hypothesis is true, one would expect to find similar distributions of farm sizes and slaveholdings in both the newer and older regions. Or, if the large planter truly led the migration, larger farms and larger slaveholdings should be a greater percentage of farms in the newer region than in the old.

It is indeed possible that the group which leads the migration into the newer region makes no contribution to the economy of the region. That is, even if we were to find that small farmers rather than large planters led the migration, one could raise the question: "But, were these smaller farmers really a part of the economy or were there just a great many of them in the region?" One way of approaching this problem is to examine the percentage of wealth holdings and crop outputs accounted for by the various farm size and slaveholding classes. If the smaller farmers or slaveholders hold a larger percentage of the wealth, produce a larger percentage of the crop output, and are numerically a greater percentage of farms in the newer region than in the older region, the validity of the hypothesis becomes doubtful.

The second part of the hypothesis will be tested by comparing the above distributions for the two census years. If the smaller farmers are being pushed out by the larger planters, large planters should own a greater percentage of the farms and produce a larger proportion of the crops in 1860 than in 1850 especially in the "better" lands of the Southwest. Because the 1850 sample is more limited in size and scope than the 1860 sample, a subsample of the 1860 sample will be used for this comparison. If the role of the yeoman farmer is more important, relative to other farmers, in the New South in 1850 than in the New South in 1860 and, at the same time, his relative importance in the Old South in 1850 is less than in the Old South in 1860, we can conclude that the yeoman farmer was not a newcomer. This would suggest that the westward expansion is best described as a movement of small farmers into the new lands followed, perhaps, by larger planters at a later date. If the relative importance of the yeoman farmer should be greater in both the New and Old South in 1850 than in 1860, any conclusions as to his role in the westward expansion must be based upon the degree of his importance in the two regions. Additionally, if his importance is less in 1860, one might conclude that the yeoman farmer had been "squeezed out" by the larger planters during the decade of the fifties. An alternative conclusion would be that the yeoman farmer had found prosperity in the newer lands.[32] If his relative importance is greater in 1860 than in 1850, we again must base any conclusions regarding the westward expansion upon degrees of importance in the regions. However, in this latter case, we would have uncovered evidence that the cotton economy was becoming more egalitarian.

The Role of the Small Farmer

As we shall see later, the manuscript census samples employed in this study provide a unique opportunity to evaluate the role of the smaller farmer and, especially, the nonslaveholding farmer in both 1850 and 1860. This analysis sheds new light on three interesting areas. First, it illuminates the role of the smaller farmer using the most extensive record which he has left us--the manuscript censuses. Second, the small farmer's contribution to the economies of each of the

---

[32] A study of this possibility utilizing a matched sample of identical farms from both the 1850 and 1860 censuses has been suggested by Professor Paul David and will be the subject of future research.

cotton-producing regions may help to explain his role in the westward expansion of cotton production. Finally, the position of the small farmer in the southern economy combined with his relative importance to that economy should provide new information on the expansionary characteristics of slavery.

Limitations of this Study

The present study has both a spatial and a temporal limitation which the reader should bear in mind. Spatially, the study is based on manuscript census data selected from the primary cotton producing regions of the South. (These regions are described in detail in Chapters II and III.) Consequently, the analysis and conclusions are not necessarily applicable to regions where rice, sugar, or tobacco were the important crops. Also, because the samples are selected from the manuscript censuses of agriculture, urban areas are omitted from consideration.

Temporally, this study is limited to consideration of data for the 1850 and 1860 census years. Any analysis involving changes over time will be confined to the single decade at the end of the antebellum period. Further, southwestern migration during the first half of the nineteenth century will be evaluated from only two vantage points both of which lie at the end of the period.

This temporal limitation is not so severe as it appears. First, as Table 1 indicates, the decades immediately preceding the 1850 and 1860 census years were ones of extremely large population increases in the newer states. The southwestern states as a group more than doubled their population during this twenty year period. In the newest states of Arkansas and Texas, population more than doubled in the last decade alone. Thus, although the 1850 and 1860 census years fall at the end of the

TABLE 1

POPULATION OF ELEVEN SOUTHERN STATES 1800 THROUGH 1860

States	1800	1810	1820	1830	1840	1850	1860
Old South							
Florida	----	----	----	34,730	54,477	87,445	140,424
Georgia	162,686	252,433	340,989	516,823	691,392	906,185	1,057,286
North Carolina	478,103	555,500	638,829	737,987	753,419	869,039	992,622
South Carolina	345,591	415,115	502,741	581,185	594,398	668,507	703,708
Virginia	886,149	983,152	1,075,069	1,220,978	1,249,764	1,421,661	1,596,318
Total	1,872,529	2,206,200	2,557,628	3,091,703	3,343,450	3,952,837	4,490,358
Decade Increase		333,671	351,428	534,075	251,747	609,387	537,521
(Percent)		(17.8%)	(15.9%)	(20.9%)	(8.1%)	(18.2%)	(13.6%)
New South							
Alabama	1,250	9,046	127,901	309,527	590,756	771,623	964,201
Arkansas	----	1,062	14,273	30,388	97,574	209,897	435,450
Louisiana	----	76,556	153,407	215,739	352,411	517,762	708,002
Mississippi	7,600	31,306	75,448	136,621	375,651	606,526	791,305
Tennessee	105,602	261,727	422,823	681,904	829,210	1,002,717	1,109,801
Texas	----	----	----	----	----	212,592	604,215
Total	114,452	379,697	793,852	1,374,179	2,245,602	3,321,117	4,612,974
Decade Increase		265,245	414,155	580,327	871,423	1,075,515*	1,291,857
(Percent)		(231.8%)	(109.1%)	(73.1%)	(63.4%)	(47.9%)*	(38.9%)

* Includes addition of Texas.

Source: U.S. Bureau of the Census, Historical Statistics of the United States, Colonial Times to 1957, Series A123-180, p. 13.

antebellum period, they are years during which southwestward migration was still very great.[33] Second, much of the analysis which follows attempts to determine the participants in southwestward migration (a flow) by examining the members of the southwestern cotton-growing economy (a stock at the end of the migration pipeline). Thus, our vantage point near the end of the migration is not particularly troublesome. Third, one portion of the study focuses on the "squeezing-out" of the smaller farmers by the larger planters. This "squeeze" is presumed to take place during the later stages of migration. Our samples, selected from the final two antebellum censuses, should put us in a perfect position from which to view this squeeze.

STIMULUS FOR THE STUDY

While engaged in selecting the Yale-UNC sample, the tremendous variety of opportunities to exploit the manuscript census data became apparent. One of the earlier attempts to apply the sample was a test of the slave productivity estimates utilized by Alfred H. Conrad and John R. Meyer, Edward Saraydar, and Richard Sutch in their recent debate

---

[33]Additional confirmation that the 1840-1860 period was one of substantial migration into the Southwest may be gained by examining the frontier lines of 1830 and 1850 as given in Frederic L. Paxson, *History of the American Frontier 1763-1893* (Boston: Houghton Mifflin Company, 1924), pp. 261 and 430. In 1830 the frontier line (six inhabitants per square mile) runs through Alabama. (The southern Mississippi delta region does have a greater population density.) By 1850, this frontier line had shifted to include all of Alabama and Mississippi, most of Louisiana and parts of Arkansas and Texas.

regarding the profitability of antebellum slavery.[34] Professor Sutch has estimated the productivity of slave labor (bales of cotton produced per slave) by dividing the cotton output for various counties by the total number of rural slaves plus farm laborers in those same counties. An implicit assumption of Professor Sutch's procedure is that all cotton was produced by slave labor or persons who listed their occupation in the census returns as "farm laborer".

Careful examination of the manuscript census reveals that a large part of the cotton crop was produced on small farms with few or no slaves. The free persons living on these farms did not report their occupations as farm laborers despite the fact that on these smaller farms the free persons worked in the fields alongside their slaves if, indeed, they owned any. Thus, Professor Sutch's estimates likely overstate the actual slave productivities.

Using the 1860 census sample (the Yale-UNC sample), it is possible to refine Professor Sutch's estimates. Rather than divide total cotton output by total rural slaves, total cotton production on farms with

---

[34] Alfred H. Conrad and John R. Meyer, "The Economics of Slavery in the Antebellum South," Journal of Political Economy, LXVI (April, 1958), pp. 95-122; Edward Saraydar, "A Note on the Profitability of Antebellum Slavery," The Southern Economic Journal, XXX (April, 1964), pp. 325-32; Richard Sutch, "The Profitability of Ante Bellum Slavery: Comment," The Southern Economic Journal, XXXI (April, 1965), pp. 365-377. A paper presenting productivity estimates based on the 1860 census sample is being prepared at the present time by James D. Foust and Dale E. Swan and will be entitled "Productivity of Antebellum Slave Labor--A Micro Approach."

slaves is divided by the total number of slaves on those farms plus the total number of free persons who likely worked alongside the slaves.[35]

These refined estimates yield a rather curious result. In the New South regions Professor Sutch's estimates indeed overstate the slave productivities. In the Old South both techniques yield approximately the same bale per slave estimates. Thus it would appear that free persons were a relatively more important part of the cotton producing labor force in the New South than in the Old South. Assuming the 1860 (Yale-UNC) sample to be representative for the moment, this result could have occurred for two reasons. First, a greater percentage of the New South farms were slaveless resulting in an overstatement of Professor Sutch's numerator. Second, the New South contained a relatively greater number of small farmers who worked alongside their slaves resulting in an understatement of Professor Sutch's denominator.

The first possibility is the easier to check. The ratio of nonslave farms to total farms can be calculated for each of the eleven states represented in the sample. The ratios are given, by state, in Table 2. The first group of states in Table 2 (Virginia, North Carolina, South Carolina, Georgia, and Louisiana) consists of the states which were settled initially. The second group (Florida, Alabama, Mississippi and

---

[35]Actually the denominator is:

slaves + 2 (explicit farm laborers + implicit farm laborers). Explicit farm laborers are those free persons who listed their occupation in the manuscript census as "farm laborer". Implicit farm laborers are defined as free males between the ages of fifteen and sixty-four who lived on farms having less than ten slaves. The factor of two is required to make both parts of the denominator comparable. Without this factor we would be adding a total population variable (slaves) to a labor force variable (farm laborers). Using a factor of two assumes a participation rate of .50.

Tennessee) were settled somewhat later. The final two states, Arkansas and Texas, were still being settled in 1860. The percentage of slaveless farms was lowest in the older states--41 percent of all farms--and highest in the frontier states--63 percent. In the middle states the figure was 54 percent.[36]

TABLE 2

PERCENTAGE OF SLAVELESS FARMS IN COTTON REGIONS OF ELEVEN SOUTHERN STATES - 1860

State	Number of Farms in Sample	Slaveless Farms in Sample	Percentage of Farms Without Slaves
Virginia	42	16	38%
North Carolina	392	191	49
South Carolina	516	192	37
Georgia	813	324	40
Louisiana	226	90	40
Total	1989	813	41
Florida	64	32	50
Alabama	1001	581	58
Mississippi	706	330	47
Tennessee	464	257	55
Total	2235	1200	54
Arkansas	434	306	71
Texas	551	314	57
Total	985	620	63

Source: Calculated from the 1860 manuscript census sample.

[36]If Louisiana is grouped with the New South and Florida with the Old South, as will be the case during later analysis, the percentages become:

Va., N.C., S.C., Ga., Fla., -- 41 percent
Ala., Miss., La., Tenn., -- 53 percent
Ark., Texas -- 63 percent

Thus, we confirm the assumption that there was a greater percentage of slaveless farms in the New South than in the Old South in 1860.

This finding is the stimulus for the present study. If slaveless farms were relatively more important in the newer regions, it would seem logical to conclude that slaveless farmers played an important role in the westward movement. If one is willing to assume a continuum of slaveholders ranging from nonslaveholders to slaveholders operating very large plantations, it would also seem logical to conclude that small slaveholders took an important part in westward expansion. Thus, the role of the yeoman farmer in westward expansion in the U.S. South appears to be more important than is generally described in the literature. The present study seeks to examine the yeoman farmer's role and, if warranted, give him the credit which he has thus far been denied.

OUTLINE OF THE PRESENT STUDY

As indicated above, the present study is based upon two statistical samples selected from the manuscript censuses of 1850 and 1860. In Chapter II, we shall describe the 1860 sample--the universe from which it is drawn, the information which was collected, the sampling procedure employed, the size and scope of the sample, and the tests used to determine the representativeness of the sample. In Chapter III, we shall discuss the 1850 sample again emphasizing the universe from which it is drawn, its size, the procedure employed, and the tests of its representativeness.

Before analyzing the data contained in the manuscript census samples, we should, perhaps, point out the need for these samples. This need is most clearly indicated by pushing the analysis so far as is possible using data available in the published census volumes. This analysis and the limitations inherent in using data from the published census will be the subject of Chapter IV.

If the first part of the hypothesis were true, i.e., if large planters led the migration, one would expect to find relatively more large planters in the Southwest than in the Southeast. In Chapters V and VI, the first part of the hypothesis will be tested by examining the distribution of farm sizes and slaveholdings in various regions in the census years 1850 and 1860 using the data in the manuscript census samples.

The second part of our hypothesis suggests that, over time, the small farmer was being pushed out of the better regions by the large planters. To test this portion of the hypothesis, the role of the small farmer in 1850 and 1860 will be contrasted. This comparison, for various regions of the South, will be made in Chapter VII.

The final analysis in this study, determining the role of the small farmer and nonslaveholder within the various regions of the South, will be the subject of Chapter VIII. Once the yeoman's role within each of the regions is ascertained, we may be better able to explain his role in the migration between regions.

The results of this study will be summarized in Chapter IX. This final chapter will also contain a brief summary of the analysis and indicate the contributions which the author has hoped to make.

CHAPTER II

THE 1860 CENSUS SAMPLE

The 1860 manuscript census consists of six schedules three of which were used in selecting the present sample. Schedule No. 1, relating to free inhabitants as of June 1, 1860, records the name, age, sex, color, value of real estate owned, value of personal estate owned, place of birth, and the occupation of each person over fifteen years old. Schedule No. 2, relating to slave inhabitants, includes the names of slave owners and the age, sex and color of each slave owned. Schedule No. 4, relating to the production of agriculture during the year ending June 1, 1860, lists the name of the owner, agent or manager; number of acres of improved and unimproved land; cash value of farm; value of farming implements and machinery; number of livestock on hand June 1, 1860; value of livestock; the quantity produced during the year of each of thirty-one crops or farm products; value of orchard products; value of the produce of market gardens; value of homemade manufactures; and value of animals slaughtered.[1]

In order to reconstruct an individual farm as it appeared to the census enumerator in 1860, it is necessary to combine the farm's free population data from Schedule No. 1, its slave population data from Schedule No. 2, and its agricultural production data from Schedule No. 4.

---

[1]For a more complete description of the Eighth Census (1860) see Carroll D. Wright, <u>History and Growth of the United States Census</u> (Washington: Government Printing Office, 1900), pp. 39-52.

Fortunately, it is possible to effect such a combination by matching the name of an individual farm owner with a head of household in the population census and a slaveowner in the slave census.[2] This matching process is very laborious. Consequently, only a limited sample of farms in the 1860 manuscript census have been matched. The sampling procedure, the resulting sample, and tests of the quality of the sample are described below.

THE UNIVERSE

It is the purpose of this study to focus attention upon the cotton growing regions of the "new" and "old" South. Therefore, the universe from which the sample is drawn is the "cotton economy" rather than the "southern economy." Limiting the sample to the "cotton economy" increases the homogeneity of the sample by excluding many areas where sugar, rice or tobacco is the dominant crop.

Three requirements are considered in determining which counties should be included in the "cotton economy." First, the "cotton economy" should indeed be a cotton producing region. Regions specializing in other crops should be excluded whenever possible. Second, a wide range of cotton producing experience should be included in the region. Large and small farms, cotton intensive farms and those producing little cotton, large slaveholdings and family farms should be included in the "cotton economy." Third, farms which produced no cotton but supplied the economy

---

[2]In many cases, it is impossible to locate the name of an individual farm owner in the slave census. This can generally be taken to infer that the individual did not own any slaves. At times, however, failure to locate a name in the slave census indicates that the slave census is incomplete or poorly taken. Value judgements must constantly be made to insure an accurate match.

with foodstuffs, slaves, etc. should be included in the economy. Hence, the "cotton economy" should be a contiguous region within which internal trade was feasible and cotton was the major cash crop.

For the purposes of this study, the "cotton economy" is defined as the region consisting of those counties producing 1,000 or more 400 pound bales of cotton in the year ending June 1, 1860. County boundaries are adhered to because the manuscript census was generally taken by county rather than by township making it impossible to distinguish subdivisions smaller than the county. The counties making up the "cotton economy" are hereafter referred to as "cotton counties."

How well does the region so defined meet these requirements? This region omits sixty percent of United States rice production in 1860, sixty-eight percent of cane sugar production, fifty-eight percent of cane molasses production, and over ninety percent of southern tobacco production. Thus the region omits areas specializing in other cash crops. Second, 409 counties producing over ninety-eight percent of total United States cotton are included in the region.[3] Individual county cotton production ranges from 1,014 bales (Sussex County, Virginia) to 141,493 bales (Tensas County, Louisiana). Finally, the 409 "cotton counties" form a generally contiguous region, limited by geography and climate, which can be considered as an economy rather than as a mere collection of counties.

The matched sample of farms from the 1860 manuscript census which is the basis of this study has been drawn from this cotton economy. That is, each of the farms in the sample was located in a county producing at least 1,000 bales of cotton in 1860. Since the Missouri manuscript census

---

[3]See Appendix I for individual state cotton production and the cotton produced in counties included in the sample.

microfilms were unavailable, the four Missouri "cotton counties" were not sampled. This omission has a negligible effect on the sample. Missouri's four counties produced approximately 38,000 bales of cotton (0.7 percent of total U. S. production) and contained less than 2,100 farms. If Missouri had been sampled, only thirty-five or forty additional farms would have been selected. Thus, the universe from which the sample has been drawn consists of the 405 remaining cotton counties. These counties are listed in Appendix I.

The universe contains approximately 263,500 farms with three improved acres or more.[4] The published census reports the number of farms with three or more improved acres by county but fails to report the total number of farms in each county. However, the total number of farms in each state is reported. In order to estimate the total number of farms in the "cotton counties" it is necessary to estimate the number of farms with less than three acres and add this estimate to the number of farms three acres and larger. This is done in one step using the following calculation for each state:

$$\text{Total number of farms} = \begin{array}{l}\text{Farms three acres}\\ \text{and larger in}\\ \text{"cotton counties"}\end{array} \times \frac{\text{Total farms in state}}{\begin{array}{l}\text{Farms three acres}\\ \text{and larger in state}\end{array}}$$

Using this procedure the total number of farms in the "cotton economy" is estimated to be approximately 300,000. The approximate number of farms in the "cotton economy," by state, is included in Table 3.

---

[4]Data on numbers of farms and numbers of farms three acres and larger are taken from U. S. Bureau of the Census, <u>Eighth Census of the United States: 1860, Agriculture</u>.

TABLE 3

NUMBER OF FARMS IN THE "COTTON ECONOMY" BY STATES

State	State Cotton Production (1,000 Bales)	Cotton Counties (Counties Producing 1,000 Bales or More)					
		No.	Cotton Production 1,000 Bales	Farms 3 Acres and Up	Adjust. Factor[a]	All Farms[b]	% of Total
Alabama	990.0	50	989.2	49,357	1.101	54,342	18.2
Arkansas	367.4	36	363.1	21,781	1.175	25,593	8.6
Florida	65.2	9	57.4	3,243	1.027	3,331	1.1
Georgia	701.8	86	683.9	38,678	1.150	44,480	14.9
Louisiana	777.7	32	774.3	13,979	1.003	14,021	4.7
Mississippi	1,202.5	53	1,196.9	36,230	1.158	41,954	14.0
Missouri[c]	41.2	4	38.2	1,994	1.048	2,090	.7
North Carolina	145.5	26	131.6	21,201	1.122	23,788	8.0
South Carolina	353.4	27	352.0	26,228	1.166	30,582	10.2
Tennessee	296.5	19	287.4	23,536	1.059	24,925	8.4
Texas	431.5	62	425.0	27,088	1.148	31,097	10.4
Virginia	12.7	5	10.8	2,171	1.071	2,325	.8
Total	5385.4	409	5309.8	265,486		298,528	100.0

[a] Adjustment Factor = $\frac{\text{Total farms in state}}{\text{Farms three acres and larger in state}}$

[b] Farms three acres and larger times the adjustment factor

[c] Missouri's four "cotton counties" were not sampled due to manuscript census microfilm being unavailable.

Source: Calculated from U.S. Bureau of the Census, Eighth Census of the United States: 1860, Agriculture

THE SAMPLE

The 1860 sample is a matched sample of farms selected from the manuscript census returns of the 405 "cotton counties." The data for each farm in the sample consist of the complete agricultural data (from Schedule No. 4), the age and sex of all persons living on the farm (from Schedule No. 1), and the age and sex of all slaves owned by the farm owner (from Schedule No. 2). Thus, drop outputs and labor inputs are recorded for each of the farms. Additionally, improved and unimproved acreage, cash value of the farm, cash value of farming implements and machinery, and the value of the owner's personal estate are available and have been recorded for each farm. These latter data at least approximate the land and capital inputs employed on each farm. The data that were recorded for each farm are listed in Appendix II.

Sampling Procedure

The matching procedure, specifically the problem of locating the farm owner's name in the free population census and again in the slave census, requires that the sample be taken in blocks of farms rather than one farm at a time. A pilot study consisting of a very few forty-farm blocks selected at wide intervals proved unsatisfactory. Due to the large intervals, more than half the cotton counties in the pilot universe were omitted from the sample and those counties which were included were represented by a single forty-farm block. Statistical tests indicated that this pilot sample was not representative of the universe.

The present sample was selected using five-farm blocks from every 280 farms in the manuscript census.[5] This is approximately a 1.67 percent sample and results in a sample size of 5,229 farms. Of course, only farms located in "cotton counties" were included in the sample.

By selecting a systematic sample, that is, selecting a five-farm block from every fourteenth page of the manuscript census, the sample is stratified by the number of farms in each of the "cotton counties." "Cotton counties" containing more farms have more pages in the manuscript census and are more frequently represented in the sample. The same principle applies to states, so the states having more farms in the "cotton economy" are more heavily represented in the sample.

Selecting blocks of only five farms assures that most counties are represented in the sample and, in most cases, each county is represented by more than one block of farms. On the average, 2.75 blocks were drawn from each "cotton county."

For each state the following sampling procedure was used:

1. The agricultural census was examined noting the number of pages of data, the quality of the data and the approximate number of farms in each county.
2. The sample was selected as follows:

    a) A random number was used to select a starting page between pages one and fourteen (sheets one and seven) of the initial county.

---

[5]The 1860 manuscript census of agriculture lists the data for forty farms on each page and requires two pages (both sides of the same sheet of paper) to record all the data for the forty farms. The sample consists of blocks of five farms selected from every fourteenth page (every seventh sheet). The exact location of the block to be selected was determined using a random number table.

b) Starting with that page, every fourteenth page (every seventh sheet) was selected. Whenever a page contained less than twenty farms it was omitted from the count.

c) The initial farm in each block was selected using a table of random numbers.

3. The name of the owner and acreage data for the five farm block plus the two farms immediately preceding the block and the two farms immediately following the block were recorded on a worksheet. The four extra farms were recorded to facilitate the matching process. Also, these extra farms served as replacements for any farms in the block which could not be matched satisfactorily.

   The matching process involved locating the name of the farm owner or operator in the free and slave population censuses. The age and sex of each of the free persons living with the owner, the age and sex of each of his slaves and the value of his personal estate were recorded on the worksheet.

4. Five of the nine farms on the worksheet were selected for inclusion in the sample. The intended five-farm block was selected if the data appeared complete and the matching satisfactory. If any farm in the block was unsatisfactory it was replaced by one of the farms preceding or following the block. In the event that five satisfactory farms could not be found on the worksheet, only those which were satisfactory were used.

5. Data from the worksheet and the crop data from the agricultural census were recorded on paper tape using a ten-key adding machine.

Two check totals were inserted at this step for later use in checking the data.[6]

6. The data on the adding machine tapes were punched into IBM cards. Perfect accuracy of this transfer was obtained using the check totals which were inserted at step five. The data on punched cards were put into the computer which then ran three checks on the data: (1) the sum of the first fifty-three items was compared to the fifty-fourth item (a check total); (2) the sum of the remaining items was compared to the second check total; (3) the total number of people on the farm was compared to the total population figure included in the data. If any of these numbers did not check, the computer printed out the data for the offending farm. The incorrect cards were corrected and the process repeated until the computer was able to go through all the farms without finding any errors.
7. Finally, the data were transferred from the corrected set of cards to magnetic tape and the above error detecting routine repeated to insure accurate transfer from cards to tape.

Size and Scope of the Matched Sample

The total sample includes data for 5229 farms in eleven states. Of the 409 "cotton counties," 382 counties (producing approximately ninety-three percent of the 1860 cotton crop) are represented in the sample.

---

[6]To insure accuracy in the transfer of data from the worksheets and census microfilm to the adding machine tape, approximately twenty percent of the farms were checked. Whenever errors were discovered on more than five percent of the farms (as when a new person was operating the adding machine), a one hundred percent check was initiated. (It should be recalled that each farm consists of at least sixty items of data. Thus, an error on five percent of the farms is an error of less than one-tenth of one percent of the data.)

Four of the omitted counties are in Missouri for which data is not available, five are too small to be included in the sample; fifteen have incomplete returns; three are omitted due to matching difficulties. The thorough coverage of the "cotton counties" and the 5229 farm sample size attest to the adequacy of the sample.

Incomplete data and matching difficulties, including the omission of twenty-seven entire counties, reduce the sample size by 426 farms (eight percent of the intended sample size of 5655 farms).

Several farms listed in the manuscript census appear to have been operated in conjunction with another farm or farms. (For example, several farms listed separately in the agricultural census were owned by members of the same household.) The sample farms include 143 farms formed by combining the crop and population data from two or more related farms. The size and scope of the sample, by state, are summarized in Appendix I.

TESTS OF THE SAMPLE

The paragraphs immediately above should convince the reader that the sample is indeed selected from the cotton producing regions of the South. The systematic sampling technique employed assures that every region and every farm had its chance to be included in the sample. Consequently, the sample should be a representative sample of farms in the "cotton counties." That is, the sample farms should be the same sizes, produce the same crops, and employ the same type of labor force as do the farms in the universe.

Statistical tests are available which will evaluate the representativeness of the sample. Several of these tests have been used to compare the sample farms to the published data for "cotton counties." The same tests have been applied using published data for counties actually included in the sample. The sample has been tested for representativeness as to crop and property data, size distribution of farms, size distribution of slaveholdings, and distribution of the population by age and sex. The samples from each of the eleven states have been tested individually and the results of these tests are included in Appendices III, IV and V. Tests of the combined sample are discussed in detail later in this chapter. Before turning to a discussion of the results of the tests of the combined sample, each of the tests should be described in detail.

<u>Crop and Property Data</u>

Two different tests have been run to compare the crop outputs[7] of the sample farms with those of all farms in the cotton counties. The first of these tests compares the total output of each crop produced on the sample farms to the amount produced in the "cotton counties." The sample includes approximately 1.7 percent of the cotton county farms. If the sample is representative of the cotton county farms; the acreage, value, and crop outputs of sample farms as a percentage of cotton county farms should be approximately 1.7 percent. These percentages for each crop, by state, are listed in Appendix III.

It should be pointed out that, strictly speaking, this test is not employed to determine the representativeness of the sample. Rather, this test is used to detect sampling errors which result in an unusually high

---

[7]Throughout this discussion of the tests of the sample, the terms "crop" or "crop output" refer to all of the data taken from the agriculture census and include dollar values of the farm, farm machinery and livestock; acres of improved and unimproved land; and the outputs of the various crops produced on the farm.

sample mean and a correspondingly high standard deviation.[8] High values of mean and standard deviation result from two types of sampling errors: first, data which were recorded in the wrong units, e.g., pounds of hay rather than tons; second, recording errors such as misplaced decimals or interchanging the output of two or more crops. Consequently, this test is able to suggest possible errors in the data but provides no indication of the representativeness of the sample. As a result of this test, however, several crop data have been examined and a few sampling errors corrected.

The second test of the crop output data provides statistical measures of the representativeness of the sample. This test compares the average output of each crop per sample farm to the average output of that crop per cotton county farm. This statistic is described below.

Let x_{ij} = output of the ith crop for the jth farm in the sample.

n = number of farms in the sample.

X_i = total output of the ith crop in the cotton counties.[9]

N = number of farms in the cotton counties (This is calculated from published census figures using:

$$\text{Farms 3 acres and up in cotton counties} \times \frac{\text{Total Farms in State}}{\text{Farms 3 acres and up in state}}$$

Then:

$$\bar{X}_i = \frac{\sum_{j=1}^{n} x_{ij}}{n} = \text{average output of the } i^{th} \text{ crop per sample farm.}$$

---

[8]The statistic, described below, which is used to indicate the representativeness of the sample will not detect a sampling error which results in a sample mean different from the population mean and a high standard deviation. This is a problem with all statistics which use a sample estimate of the standard deviation as the basis of comparing a sample static with some population parameter.

[9]The same statistic is also calculated using universe data taken from the counties actually represented in the sample. Either method yields approximately the same results. (See Appendix IV.)

$$\mu_i = \frac{X_i}{N}$$ = average output of the ith crop per cotton county farm.

$$S_i = \sqrt{\frac{\sum_{j=1}^{n} (X_{ij} - \bar{X}_i)^2}{n-1}}$$ = estimate of the standard error of the crop output for the ith crop.

$$S_{\bar{X}_i} = \frac{S_i}{\sqrt{n}}$$ = estimate of the standard error of the mean for the ith crop.

$$t_i = \frac{\bar{X}_i - \mu_i}{S_{\bar{X}_i}}$$ = random variable which is used to measure the divergence of the sample mean from the cotton county mean for the ith crop.

Each of the t_i's thus obtained has a probability distribution which is a t-distribution with (n-1) degree of freedom. (Actually for a sample size greater than one hundred the distribution is approximately a normal distribution with a mean of zero and a variance of one, i.e., N(0,1). That is, if the sample is randomly drawn from the cotton counties, the probability of getting a t_i which is greater than +2 or less than -2 is only .0455.)

This statistic is used to determine the representativeness of the sample with respect to each crop. The t_i values for each crop, by state, are presented in Appendix IV.

Size Distribution of Farms

The sample is intended to represent all sizes and types of farms in the cotton counties. If the sample is indeed representative of all sizes of farms, the percentage of sample farms in each size category should be approximately the same as the percentage in the cotton counties. The size distribution of farms three acres and larger in each county is printed in the published census.

A chi-square statistic is used to determine how well the sample distribution approximates the cotton county distribution. Let:

k = number of size classifications used.

n' = number of farms in the sample three acres and larger.

N' = number of farms in the cotton counties three acres and larger.[10]

e_i = number of cotton county farms in the ith size classification.

O_i = number of sample farms in the ith size classification.

Then

$\frac{e_i}{N'}$ = the proportion of cotton county farms in the ith size classification.

and $E_i = \frac{e_i}{N'} n'$ = the expected number of sample farms in the ith size classification.

The statistic used is:

$$X^2 = \sum_{i=1}^{k} \frac{(O_i - E_i)^2}{E_i}$$

The sampling distribution of this statistic is approximately a chi-square distribution with k-1 degrees of freedom. Thus, if there are seven categories, the statistic is approximately a chi-square distribution with six degrees of freedom. In order for this approximation to be close, it is necessary for the expected number of farms in each category to be greater than one and at least eighty percent of them must be greater then five. Except where categories have to be combined to satisfy this restriction, the following size categories are used:

---

[10] Again, this statistic is also calculated using universe data taken from the counties actually represented in the sample. Either method yields approximately the same results. For a comparison of the results see Appendix V.

3- 9 acres
10- 19 acres
20- 49 acres
50- 99 acres
100- 499 acres
500- 999 acres
1000 acres and up.

Farms of less than three acres can not be included in this test as the size distribution of farms in the published census only includes farms three acres and larger.

The X^2 statistic is compared to the proper chi-square distribution to determine whether the sample farm sizes are significantly different from the size distribution of farms in the universe. By comparing the individual O_i and E_i values in each size category it is possible to determine which of the categories contain too many or too few farms.

The results of these tests, by state, are included in Appendix V. The probability levels indicate the probability of obtaining an X^2 value as large or larger than the one obtained if the sample farms were truly a random sample of the universe farms.

<u>Size Distribution of Slaveholdings</u>

To determine whether the slaveholdings on the farms in the sample were representative of cotton county slaveholdings, a statistic similar to the one immediately above is used. The only modification required is that e_i, E_i, and O_i now refer to the number of farms (or expected number of farms) in <u>slaveholdings</u> classifications rather than farm-size classifications.

The following slaveholding classifications are used except in situations where categories had to be combined in order to satisfy the requirement that all expected values must be greater than one and at least eighty percent must be greater than 5:

1	slave
2	slaves
3	slaves
4	slaves
5	slaves
6	slaves
7	slaves
8	slaves
9	slaves
10- 14	slaves
15- 19	slaves
20- 29	slaves
30- 39	slaves
40- 49	slaves
50- 69	slaves
70- 99	slaves
100- 199	slaves
200- 299	slaves
300- 499	slaves
500- 999	slaves
1000 slaves or more	

The X^2 statistic is compared to the proper chi-square distribution to determine the representativeness of the sample and the categories in which the larger deviations occur. It should be pointed out that the size distribution of slaveholdings for cotton counties is calculated from county totals in the published census and these county figures include slaveholdings of both farm and non-farm slaveowners. Although the use of this test appears to imply that the distribution of slaveholdings is the same for farm and non-farm slaveowners, no such situation is postulated.

As in the earlier tests, this statistic is calculated twice using as the universe: first, the cotton counties and, second, the counties actually represented in the sample. The results of this test, by state, are included in Appendix V. Again, the probability levels indicate the probability of obtaining an X^2 value as large or larger than the one obtained if the sample farms were truly a random sample of the universe farms.

Distribution of Population by Age, Sex and Condition

Two chi-square statistics are used to determine whether the population on the sample farm is representative of the population in the cotton counties. For the first test, the population is classified in four categories: free males, free females, slave males and slave females. The distribution of the population on the sample farms is then compared to the distribution of cotton county population other than in cities and towns[11] using the above chi-square statistic. In this case, e_i, E_i, and O_i refer to the number of people in each condition-sex category and the sampling distribution of the statistics is a chi-square distribution with three degrees of freedom.

The second test considers the age distribution of persons within each condition-sex category. The members of each condition-sex category are classified according to the following categories:

Under 5 years
5- 9 years
10-14 years
15-19 years
20-29 years
30-39 years
40-49 years
50-59 years
60 years and older

(Once again, categories are combined any time the expected value in any category is less than one or fewer than eighty percent of them are greater than five.) The four age distributions (one for each condition-sex classification) are compared to the age distributions of each condition-sex

[11]The published census contains sex and condition data for each county and also the population of cities and towns in each state by sex and condition. The distribution of cotton county population outside of cities and towns is computed by totaling the data for each of the cotton counties and then subtracting the population of each city or town in these counties.

classification in the cotton county population by computing a chi-square statistic for each of the four condition-sex classifications. For this chi-square statistic, e_i, E_i, and O_i refer to the number of persons in each of the nine age categories.

In each of the age, sex and condition tests, the cotton county data are taken from the published census. The published census distributions include both farm and non-farm people. As in the test of slaveholdings, it is <u>not</u> postulated that age, sex and condition are the same for farm and non-farm populations. Rather, the sample distributions are compared to the published distributions <u>and</u> an attempt is made to determine whether any deviations could reasonably be attributed to farm--non-farm differences.

As in the earlier tests, these statistics are calculated using both cotton counties and sampled counties as the universe. The results of these tests, by state, are included in Appendix V. Again, the probability levels indicate the probability of obtaining an X^2 value as large or larger than the one obtained if the sample farms were truly a random sample of the universe farms.

<u>Summary of the Tests</u>

1. Representativeness of the sample with respect to each of the crop and property data is tested using the statistic:

$$t = \frac{\bar{Y} - \mu}{S_{\bar{X}}}$$

2. Size distribution of farms, size distribution of slaveholdings, and the distribution of population by age, sex and condition are tested using the following chi-square statistic:

$$X^2 = \sum_i \frac{(O_i - E_i)^2}{E_i}$$

It is important to remember that these tests do not enable us to say that the sample is indeed representative of the cotton (or sampled) counties. Rather, a lack of significance in any of the statistics indicates only that there is insufficient evidence to conclude that the sample is unrepresentative.

Each of the tests above was described using data for the cotton counties as the standard of comparison. Similar tests will be run using the counties actually sampled as the standard. However, for the purposes of this paper, the most important comparison is that with the cotton county data as it is the "cotton county" universe which this paper intends mainly to consider.

RESULTS OF THE TESTS

Each of the two types of tests described above have been applied to the entire 5229 farm sample, first, using the cotton county data as the standard of comparison and, second, using the counties actually included in the sample as the standard.

Crop and Property Data

The first test of the sample is designed to determine the representativeness of the sample with respect to each of the crop and property data by comparing the mean crop outputs of the sample farms to the mean output of the farms in the universe. The second and third columns of Table 4 present the results of this test if the cotton counties are taken to be the universe. Taking the counties actually represented in the sample as universe yields the results in the fourth and fifth columns.

TABLE 4

TESTS OF THE AGGREGATE 1860 SAMPLE--t-STATISTICS

Crop	$\bar{x}$	Cotton Counties u	Cotton Counties t	Sampled Counties u	Sampled Counties t
Acres (improved)	129.85	133.28	-1.17	131.42	-.54
Acres (unimproved)	306.65	317.76	-1.04	314.46	-.73
Cash Val. of Farm	4363.52	4237.02	.62	4137.29	1.10
Val. of Imp. & Mach.	194.37	199.94	-.46	186,06	.69
Horses	3.51	3.52	-.05	3.52	-.05
Asses & Mules	2.29	2.35	-.50	2.32	-.25
Milch Cows	5.54	6.00	-4.18	5.96	-3.82
Working Oxen	2.05	1.98	1.17	2.09	-.67
Other Cattle	15.22	15.54	-.25	15.14	.06
Sheep	9.64	8.84	1.36	8.85	1.34
Swine	36.44	37.77	-2.02	37.73	-1.95
Val. of Livestock	890.81	921.60	-1.47	912.86	-1.05
Wheat	34.10	32.83	.91	33.47	.45
Rye	2.32	2.53	-.88	2.56	-1.00
Corn	631.78	651.33	-1.32	653.19	-1.44
Oats	18.79	17.57	.83	17.78	.69
Rice[a]	48.89	256.49	-12.99[a]	263.80	-13.45[a]
Tobacco	56.10	68.93	-1.58	70.99	-1.83
Cotton	18.24	20.02	-2.44	19.57	-1.82
Wool	18.36	16.62	.99	16.69	.95
Peas & Beans	36.25	36.17	.05	36.09	.09
Irish Potatoes	9.61	9.96	-.88	9.80	-.48
Sweet Potatoes	107.86	107.17	.20	107.08	.22
Val. of Orchard Prod.	6.46	5.07	1.21	5.02	1.25
Val. of Mkt. Garden Prod.	2.57	2.89	-.71	2.83	-.58
Butter	118.73	119.33	-.26	121.09	-1.01
Cheese	.30	.41	-1.10	.42	-1.20
Hay	4.24	1.49	1.87	1.47	1.88
Cane Sugar	.36	.28	.42	.19	.89
Molasses	31.06	24.70	.46	17.09	1.00
Beeswax	1.37	1.38	-.13	1.39	-.25
Honey	17.99	19.11	-1.44	19.44	-1.86
Val. of Home Manuf.	27.89	28.57	-.92	29.08	-1.61
Val. of Animals Slaught.	194.58	193.43	.16	195.43	-.12

[a]Rice data are not reliable due to census taker errors in recording South Carolina production.

Source: Calculated from 1860 manuscript census sample and U.S. Bureau of the Census, _Eighth Census of the United States: 1860, Agriculture_.

Using a five percent rejection level ($|t| > 1.96$) only four of the thirty-four sample means are significantly different from those of the cotton county farms. One of these four, rice, results from census taker errors in reporting South Carolina production. Unfortunately, both swine production and cotton production per sample farm are significantly less than their respective productions on the universe farms.

If the counties actually included in the sample are taken to be the universe, only two of the thirty-four sample means are significantly different from the universe means (using a five percent rejection level). Again, the rice value is of doubtful validity. In this case, the cotton and swine means are not significantly different from the universe values.

The above results provide insufficient grounds to conclude that the sample is not representative of the universe. In fact, if the sample were truly representative of the universe, one would expect to find approximately two of the sample means to be significantly different from the universe values. The results of the tests using the counties actually included in the sample were representative. Unfortunately, it appears that the mean crop outputs of the sample may be biased in the direction of mean values which are lower than the universe values.

Size Distribution of Farms

The second test of this sample compares the size distribution of farms in the sample to the size distribution of farms in "cotton counties" and in the counties actually included in the sample. The results of this test are presented in Table 5. Unfortunately, data regarding the number of farms of less than three acres in the cotton counties or in the sampled counties are not available. Consequently, this test can only be applied to farms three acres and larger.

TABLE 5

TESTS OF THE AGGREGATE 1860 SAMPLE
(Size Distribution of Farms)

Farm Sizes	Number of Sample Farms	Expected Number of Farms: Cotton Counties	Expected Number of Farms: Sampled Counties
3- 9 acres	108	117.0	113.1
10- 19 acres	403	427.6	422.5
20- 49 acres	1480	1497.3	1499.9
50- 99 acres	1281	1237.7	1245.1
100-499 acres	1607	1596.4	1599.3
500-999 acres	196	212.8	210.2
1000 acres and over	79	65.2	63.9
Total	5154	5154.0	5154.0
Farms less than 3 acres	75	$x^2(df = 6) = 8.14$ $p = .23$	$x^2(df = 6) = 7.00$ $p = .33$

Source: See Table 4.

Using either the cotton counties or sampled counties as universe and a five percent level of significance, the distribution of farm sizes in the sample does not significantly differ from the universe distributions. Examination of Table 5 might lead one to conclude that the sample is somewhat biased in the direction of larger farms but this sort of analysis, although useful, perhaps, cannot be supported by the statistical findings. Thus, we conclude that the size distribution of the farms in the sample is not significantly different from that of the universe.

Size Distribution of Slaveholdings

A comparison of the size distribution of slaveholdings on the sample farms with the distribution of slaveholdings in the universe constitutes the third test of the sample. Before discussing the results of this test, it should again be mentioned that the size distribution of slaveholdings in the cotton counties and the sampled counties includes both farm and non-farm slaveholdings. Thus, it would be surprising if the size distribution of slaveholdings in the sample were the same as that in the cotton counties or sample counties. (Recall that the sample consists of a collection of <u>farms</u> selected from the manuscript census of agriculture.)

The results of this test, given in Table 6, indicate that the size distribution of slaveholdings in the sample is significantly different from that of the cotton counties or sampled counties even at a .0005 level. A glance at Table 6 reveals that the sample is deficient in slaveholdings of seven slaves or less and in the nine-slave category. Of course, the sample includes an excess number of slaveholdings in each of the other categories. These results are approximately what the writer expected. One would expect any sample consisting only of farms to be deficient in the smaller slaveholding classes due to the fact that many of the smaller slaveholdings would likely be made up of domestic servants. Presumably, many domestic servants would be found in urban households. On the other hand, it is unlikely that many of the large slaveholdings would be found in urban areas.

Because the size distribution of slaveholdings in the sample is approximately what one would anticipate, the representativeness of the sample cannot be rejected on the basis of this test.

TABLE 6

TESTS OF THE AGGREGATE 1860 SAMPLE
(Size Distribution of Slaveholdings)

Number of Slaves	Number of Sample Slaveholders	Expected Number of Slaveholders: Cotton Counties	Expected Number of Slaveholders: Sampled Counties
1	294	416.2	416.7
2	219	272.2	272.9
3	160	213.0	213.5
4	151	180.5	180.8
5	145	154.7	155.0
6	113	133.4	133.8
7	102	112.3	112.6
8	125	102.2	102.6
9	84	86.9	86.8
10-14	335	300.7	301.4
15-19	237	173.4	173.5
20-29	243	183.4	183.1
30-39	153	93.0	92.4
40-49	79	52.7	51.8
50-69	67	55.0	53.9
70-99	42	34.6	33.9
100-199	37	20.9	20.3
200-299	1	2.2	2.1
300-499	1	.6	.6
500-999	--	.1	.1
1000 and over	--	--	--
Total	2588	2588.0	2587.8
Farms without slaves	2641	$x^2(df = 17) > 42.84$ $p < .0005$	$x^2(df = 17) > 42.84$ $p < .0005$

Source: See Table 4.

<u>Age Distribution of the Population</u>

The final test of the sample consists of a comparison of the age distribution of the sample population (by sex and condition) with the age distribution of the cotton county and sample county populations. Again, the universe data consist of both urban and rural populations. Thus, it would be surprising if the sample distributions were similar to the universe distributions.

The age distributions of both free males and free females in the sample counties are significantly different (at a .0005 level) from those in both the cotton counties and in the counties actually included in the sample. (See Table 7.) In both cases there is a deficit of persons aged less than five years and persons between the ages of 20 and 39. Perhaps the deficit of persons in the 20-39 age group reflects the fact that this is the group which leaves the farm to seek urban employment. Also, this age group may tend to live on farms of less than three acres and make up a rural wage-labor force. In any case, the deficit of persons under five years of age is a corollary of the shortage of persons in the 20-39 age group.

The age distribution of slave males is significantly different from the cotton county distribution at the five percent level of rejection. However, the sample distribution is not significantly different from the age distribution of slave males in the counties actually included in the sample. The similarity of these two distributions may be explained by the fact that the great majority of slaves were held in the rural areas and, undoubtedly, the major occupation of slave males consisted of farm labor.

The age distribution of slave females in the sample is significantly different (at a .0005 level) from the age distribution in both the cotton counties and the sampled counties. The sample contains an excess of slave females between the ages of 15 and 29 and a corollary excess of children (both male and female) under five years of age. This excess of slave females in the 15-29 age group probably results from the fact that most slave breeding went on in the rural areas. Actually, there is an excess of both slave males and slave females in the 15-19 age group, the age group which is generally thought of as consisting of prime field hands.

TABLE 7

TESTS OF THE AGGREGATE 1860 SAMPLE
(AGE DISTRIBUTIONS BY SEX AND CONDITION)

	Ages	Number in Sample	Expected Number: Cotton Counties	Expected Number: Sampled Counties
Free Males				
	Under 5	2507	2689.7	2693.9
	5-9	2563	2459.1	2461.3
	10-14	2470	2278.6	2284.4
	15-19	2044	1814.8	1819.1
	20-29	2628	3024.3	3016.8
	30-39	1710	1951.8	1942.3
	40-49	1356	1295.0	1291.3
	50-59	895	778.5	778.9
	60 & over	703	584.3	587.8
	Total	16876	16876.1	16875.8
			$x^2(df=8) > 27.84$ $p < .0005$	$x^2(df=8) > 27.84$ $p < .0005$
Free Females				
	Under 5	2372	2519.5	2518.1
	5-9	2360	2312.4	2312.2
	10-14	2308	2121.8	2122.9
	15-19	1959	1845.1	1845.7
	20-29	2384	2746.8	2744.3
	30-39	1628	1682.7	1680.3
	40-49	1191	1083.1	1084.3
	50-59	719	651.6	653.1
	60 & over	574	531.9	534.0
	Total	15495	15494.9	15494.9
			$x^2(df=8) > 27.84$ $p < .0005$	$x^2(df=8) > 27.84$ $p < .0005$
Slave Males				
	Under 5	3501	3376.1	3401.9
	5-9	2993	2966.6	2969.0
	10-14	2783	2845.2	2864.8
	15-19	2391	2282.4	2293.6
	20-29	3725	3767.9	3741.2
	30-39	2198	2282.2	2263.2
	40-49	1408	1432.7	1425.9
	50-59	748	782.3	777.7
	60 & over	650	661.6	659.7
		20397	20397.0	20397.0
			$x^2(df=8) = 17.11$ $p = .025$	$x^2(df=8) = 12.97$ $p = .12$

TABLE 7 (Continued)

Ages	Number in Sample	Expected Number: Cotton Counties	Expected Number: Sampled Counties
Slave Females			
Under 5	3565	3435.7	3449.6
5-9	2938	2949.2	2944.8
10-14	2627	2679.4	2695.3
15-19	2495	2374.6	2382.8
20-29	3769	3670.8	3647.2
30-39	2164	2295.0	2290.0
40-49	1338	1414.9	1412.9
50-59	688	731.5	729.8
60 & over	595	627.9	626.5
Total	20179	20179.0	20178.9
		x^2(df=8) = 30.63 $p < .0005$	x^2(df=8) = 29.83 $p < .0005$

Source: See Table 4.

Although the above rationalizations of the discrepancy of the sample distributions and the universe distributions may be somewhat questionable, there seems to be little grounds for rejecting the sample on the basis of this test.

Summary

As mentioned earlier, the above tests of the sample do not permit us to determine that the sample is indeed representative of the universe. Rather, the tests only suggest whether there is sufficient evidence to reject the sample as being unrepresentative. In each of the four tests applied, we are unable to determine that the sample is significantly different from the universe. Consequently, we are unable to find satisfactory grounds for rejecting the sample. By default, we will accept it as being representative of the universe under consideration, i.e., the 405 counties which produced more than 1000 bales of cotton in the 1859 production year.

CHAPTER III

THE 1850 CENSUS SAMPLE

The 1860 sample, described in the previous chapter, is part of a project that is much larger in scope than the present paper. Consequently, the 1860 sample encompasses all the cotton growing regions and includes complete manuscript census data.

In order to extend the analysis in this paper, a limited sample was selected from the 1850 manuscript census. This sample includes data from a smaller geographic region than the 1860 sample and contains only a portion of the available manuscript census data.

The 1850 manuscript census is similar to the 1860 census described in the previous chapter with three exceptions: (1) only occupations of males over 15 years of age were returned; (2) value of personal estate was not returned; and (3) the quantity of honey and beeswax was combined in one entry.

THE UNIVERSE

The 1850 sample is intended to supplement the larger 1860 sample and is confined to a more limited geographic region. The counties selected are those cited by Gray as the major cotton producing counties of 1859.[1] They are separated into three groups--alluvial regions, other New South regions, and Old South regions--according to soil and climatic conditions.

---

[1] Lewis Cecil Gray, History of Agriculture in the Southern United States to 1860, 2nd ed. (Gloucester, Mass.: Peter Smith, 1958), pp. 531-37.

Seventy-four counties are included in these three regions. These counties, listed by region in Appendix VI, are the universe from which the 1850 sample is selected.

The universe contains approximately 46,300 farms.[2] These farms produced 905,608 bales of cotton in 1850, approximately 37 percent of the United States total.

THE SAMPLE

The 1850 sample is a matched sample of farms selected from the manuscript census returns of the seventy-four universe counties. The agricultural data for each farm in the sample include improved and unimproved acreage; value of farm; value of farm machinery; number of swine; and corn, rice, tobacco, cotton, and sweet potato production.[3] Free persons living on each of the sample farms are classified, by sex, in three age categories: under 15 years, 15 thru 64 years, and over 65 years. The slave population is recorded using the same categories. Value of real estate, included in the population census, is also recorded. This latter item is generally identical to the value of the farm listed in the agricultural census.

Sampling Procedure

As discussed in the previous chapter, the matching procedure requires that the sample be taken in blocks of farms rather than one farm at a time.

---

[2]Data on numbers of farms and cotton output were taken from U.S. Bureau of the Census, *Seventh Census of the United States: 1850, Compendium.*

[3]Initially, data were recorded for sugar and molasses production. These data were discarded once it became clear that sugar was an insignificant crop in each of the regions.

The 1850 sample was selected using three farm blocks. In the Alluvial region, a three-farm block was selected from every 82 farms in the manuscript census. This is approximately a 3.66 percent sample and results in a sample size of 365 alluvial farms. In the New South region, a three-farm block was selected from every 164 farms (a 1.83 percent sample). In the Old South region, the block was selected from every 205 farms (a 1.46 percent sample). The resulting sample contains 223 farms from the New South region and 309 farms from the Old South.[4]

The reliability of a sample is a function of the absolute size of the sample. Different sampling procedures were used for each region to insure that an adequate number of farms were selected from each region. So long as the percentage of farms included in the sample remains low, say less than ten percent, the percent sampled has little effect on sample reliability.[5]

By selecting a systematic sample, the sample is stratified by the number of farms in each of the counties. Counties containing more farms have more pages in the manuscript census and are more frequently represented in the sample.

---

[4]The 1850 manuscript census of agriculture lists the data for forty-one farms on each page and requires two pages (both sides of the same sheet of paper) to record the complete data for forty-one farms. The sample consists of blocks of three farms selected from every fourth page (every second sheet) in the Alluvial region, every eighth page (every fourth sheet) in the New South region, and every tenth page (every fifth sheet) in the Old South region. The exact location of the block to be selected was determined using a table of random numbers.

[5]For a discussion of this point see Frederick E. Croxton and Dudley J. Cowden, Practical Business Statistics (Englewood Cliffs: Prentice-Hall, Inc., 1960), pp. 173-174.

The precise sampling procedure used in selecting the 1850 sample is similar to that used for the 1860 sample. The details of this procedure are included in the previous chapter.

Size and Scope of the Matched Sample

The total 1850 sample includes 897 farms selected from sixty-three counties in seven states. Of Gray's seventy-four major cotton producing counties, only eleven are omitted from the sample. Three of these were not counties in 1850; five Georgia counties were omitted due to the loss of one roll of microfilm; two counties had poor quality data; and one county was omitted due to matching difficulties.

Incomplete data and matching difficulties, including the omission of eight entire counties, reduce the sample size by 123 farms (twelve percent of the intended sample size of 1020 farms). Unfortunately, seventy-eight of the omitted farms are in the Old South region.

Of the 897 farms in the sample, 309 are selected from the Old South region, 365 from the Alluvial region, and 223 from the New South region. A summary of the number of farms in the 1850 sample by county, state, and region is included in Appendix VI.

TESTS OF THE SAMPLE

As in the case of the 1860 sample, the systematic sampling technique employed in selecting the 1850 sample assures that every farm in the universe had its chance to be included in the sample. Consequently, the sample should be representative of the farms in the three major regions--Old South, Alluvial, and New South--and the appropriate subsamples should be representative of each of the three regions.

The published census for 1850 lists crop outputs and the age distribution of the population by county. However, the size distribution of farms and size distribution of slaveholdings were not calculated. Consequently, it is possible to apply only two of the representativeness tests to the 1850 sample. The first test which can be used is a comparison of the average output of each crop per sample farm to the average output of that crop per universe farm. The statistic employed is:

$$t_i = \frac{\bar{x}_i - \mu_i}{S_{\bar{x}_i}}$$

where $\bar{x}_i$ = average output of the ith crop per sample farm

μ_i = average output of the ith crop per universe farm (Calculated by weighting the universe crop outputs and number of farms by the proportions in which they are represented in the sample. The sample consists of approximately 1.46% of all farms in the Old South region; 3.66% of all farms in the Alluvial region, and 1.83% of all farms in the New South region. Thus, in calculating the weighted u_i the Old South data, the Alluvial data and the New South data were weighted by factors of 1.00, 2.50, and 1.25 respectively).

and $S_{\bar{x}_i}$ = estimate of the standard error of the mean for the ith crop.

The other test of the 1850 sample involves a comparison of the age distribution of each of the four condition-sex groups (free-male, free-female, slave-male, slave-female) in the sample with the age distribution of each group in the universe counties. The persons in each condition-sex category on the sample farms are classified in three age categories: under fifteen, fifteen thru sixty-four, and sixty-five and over. The four age distributions (one for each condition-sex classification) are compared to the age distribution of each condition-sex classification in the universe population by computing the chi-square statistic:

$$x^2 = \sum_{i=1}^{k} \frac{(O_i - E_i)^2}{E_i}$$

where k = number of age classifications used = 3;
n' = number of persons in the particular condition-sex category in the sample;
E' = number of persons in the particular condition-sex category in the universe;
e_i = number of persons of the particular condition and sex in the ith age category in the universe;
O_i = number of persons in the sample of the particular condition and sex who are in the ith age category;

and $E_i = \frac{e_i}{N'} \cdot n'$ = the number of persons of the particular condition and sex in the sample who are expected to be in the ith age classification.

As in the case of the mean crop-output test, each of these universe values is a weighted sum of the regional values. Thus, in calculating N' and e_i the Old South data were weighted by a factor of 1.00, the Alluvial by 2.50 and the New South by 1.25. (Alternatively, of course, appropriate adjustments could have been made in the sample values.)

As mentioned in Chapter II, these tests do not permit us to determine that the sample is indeed representative of the universe. Rather, the tests are able only to suggest whether there is sufficient evidence to indicate that the sample crop outputs per farm and age distribution of the sample population are significantly different from the universe parameters. Further, the expected number of persons in each age category (utilized in the second test) is calculated using data taken from the published census. These published age data include both urban and rural populations. Nevertheless, they are the best available data and, as such, may shed some light on the representativeness of the sample.

Before considering the results of these tests, the exact universe against which the sample is tested should be defined. Recall that the universe defined earlier in the chapter consisted of the seventy-four counties designated by Gray as the major cotton producing counties in 1860. Three of these were not counties in 1850. Further, the manuscript census microfilm of five other counties has been lost.[6] Thus, for the purposes of testing the 1850 sample, the universe is defined as Gray's seventy-four counties less the three which were not counties in 1850 and the five counties for which data were unavailable.

RESULTS OF THE TESTS

Each of the tests described above was applied to the 897 farms making up the 1850 sample. Again, the universe against which the sample is tested is a weighted average of the crop, property and population data of each of the three regions. The results of these tests are described below.

Crop and Property Data

This first test of the sample is designed to evaluate the representativeness of the sample with respect to each of the crop and property data. In this test, the mean crop outputs (or property values) of the sample farms are compared to the mean crop outputs on the universe farms using the t-statistic described earlier in the present chapter. The results of this test are summarized in Table 8.

It should be recalled that only ten items of crop and property data were recorded for the more limited 1850 samples. (The 1860 sample was selected for use in a variety of projects much wider in scope than the present project. Consequently, complete crop and property data were recorded for each farm in the 1860 sample.) Using a five percent rejection level ($|t| > 1.96$) none of the ten sample means are significantly different

---

[6] Stewart, Sumter, Thomas, Troup and Washington Counties, Georgia.

from those of the universe farms. In fact, if the rejection level is increased to ten percent ($|t| > 1.645$) we are still unable to reject any one of the ten sample means as being significantly different from the universe mean.

TABLE 8

TESTS OF THE AGGREGATE 1850 SAMPLE
(Mean Crop Outputs Per Farm)

Crop	Weighted u	$\bar{x}$	$S_{\bar{x}}$	t
Improved Acres	160.36	153.19	7.41	-.97
Unimproved Acres	316.08	302.76	23.55	-.57
Cash Value of Farm	3262.24	3397.32	256.13	.53
Value of Implements & Machinery	250.63	249.81	20.44	-.04
Swine	51.61	49.75	1.79	-1.04
Corn	863.96	803.56	37.42	-1.61
Rice	54.92	52.67	12.09	-.19
Tobacco	2.32	3.27	1.18	.81
Cotton	23.46	23.55	1.76	.05
Sweet Potatoes	154.08	142.55	8.32	-1.39

Source: Calculated from the 1850 manuscript census sample and U.S. Bureau of the Census, Seventh Census of the United States: 1850.

Despite the impossibility of rejecting any of the sample means, the fact that seven of the ten t-values are negative might lead one to the conclusion that the sample is biased toward farms with crop and property data smaller than the universe average. Although this *may* indeed be the case, seven minus signs out of ten is not statistically significant. In a set of ten differences each of which has a probability distribution with median equal to zero, the probability of having seven positive values and three negative values is slightly over seventeen percent. Consequently, we cannot statistically conclude that the sample is biased toward farms having smaller than average property values and crop outputs.

<u>Age Distribution of the Population</u>

The second test of the sample consists of a comparison of the age distribution of the population on the 897 sample farms with the age distribution of the universe population. This comparison is made for each of the four condition-sex groups--free-males, free-females, slave-males, and slave-females. For this test, the universe data consists of both urban and rural population while the sample is selected from rural regions only. Thus it would be surprising if the sample distributions were similar to the universe distributions.

The age distribution of free males on the sample farms is significantly different from the age distribution of free males in the universe. Surprisingly, the age distribution of each of the other condition-sex groups is not significantly different for the sample and the universe. (See Table 9.)

Because only three age classifications were used to record the 1850 population data, it is difficult to formulate any explanation of the difference in the sample and universe age distributions of free males. The only observation possible is the obvious one--there are relatively too many free males under fifteen and too few between the ages of fifteen and sixty-four. Perhaps many men in this latter age group are members of the wage labor force, either rural or urban, and are not members of a farm household. (Hired laborers who worked on the farm but resided elsewhere would not have been included in either the 1850 or the 1860 sample.)

Presumably, one could argue that these men were less likely to be married than were individuals who owned their own land and thereby account for the fact that the relative scarcity of free-females in the fifteen thru sixty-four age group is less pronounced.

TABLE 9

TESTS OF THE AGGREGATE 1850 SAMPLE
(Age Distribution by Sex and Condition)

	Ages	Number in Sample	Expected Number
Free Males	Under 15	1337	1253.9
	15-64	1510	1593.5
	65 & Over	49	48.6
	Total	2896	2896.0
		x^2(df=2) = 9.88 p = .008	
Free Females	Under 15	1175	1153.5
	15-64	1292	1307.2
	65 & Over	40	46.3
	Total	2507	2507.0
		x^2(df=2) = 1.44 p = .49	
Slave Males	Under 15	2531	2510.3
	15-64	3128	3142.5
	65 & Over	94	100.2
	Total	5753	5753.0
		x^2(df=2) = .62 p = .70	
Slave Females	Under 15	2366	2415.8
	15-64	2996	2948.3
	65 & Over	90	87.8
		5452	5451.9
		x^2(df=2) = 1.86 p = .40	

Source: Calculated from the 1850 manuscript census sample and U.S. Bureau of the Census, Seventh Census of the United States: 1850.

The similarity of the age distributions of the sample and universe slave population is not so surprising. The great majority of slaves were held in the rural areas. Therefore, any age distribution of slaves selected from both rural and urban areas would be heavily influenced by the age distribution of the rural members. In fact, as our results indicate, the age distribution of rural and urban slaves taken together may not be significantly different from the age distribution of rural slaves taken alone.

The reader may well have some quarrel with the rationalizations of the above two paragraphs. However, these tests of the age distribution of the population provide little evidence for rejecting the sample.

Summary

As in the tests of the 1860 sample, the above tests do not permit us to determine that the sample is representative of the universe, If significant, they only suggest that there is sufficient evidence to reject the sample as being unrepresentative. In each of the tests applied, we are unable to conclude that the sample is significantly different from the universe. Thus, we have no grounds for rejecting the sample and, by default, accept it as being representative of the sixty-six county universe.

CHAPTER IV

LIMITATIONS OF PUBLISHED CENSUS DATA

Before proceeding to an analysis based on our manuscript census samples, we should first exhaust the analytical opportunities available to us using the data in the published census volumes. Unfortunately, the published data is rather limited. Size distributions of farms and size distributions of slaveholdings, vy county, were published for 1860 but not for 1850. However, size distributions of slaveholdings, by state, were published for both 1850 and 1860.

TESTING THE HYPOTHESIS

Any analysis based on the published census must obviously be quite limited. However, the 1860 size distributions of farms, by county, can be used to test our hypothesis. The 1860 size distributions of farms, by state, can also be utilized to test our hypothesis but any conclusions obtained will apply to all of the farms in the states rather than to the primary cotton-growing regions only. Consequently, any analysis based on state totals will be of questionable value. The problem here is the usual problem when heterogeneous items are compared--apples are compared to oranges. In our case, cotton growing regions would be contrasted to rice, sugar or tobacco producing regions.

The 1860 size distributions of slaveholdings can also be utilized to test our hypothesis. However, the slaveholding data in the published census includes both rural and urban slaves. Thus, any comparison of slaveholding sizes in various regions will be effected by differences in the urban slaveholdings in the regions. The 1850 and 1860 size distributions of slaveholdings, by state, can also be used to test our hypothesis but, again, any conclusions reached will apply to the entire states rather than to cotton growing regions only and rural-urban differences will cause even more distortion.

Tests of the Hypothesis--1860 Data

The first part of our hypothesis--that the migration into the new land in the Southwest was one of large planters leading, accompanying, or immediately following smaller farmers--can be tested using either 1850 or 1860 data. The first tests of this part of our hypothesis utilize farm size and slaveholding data, by county, from the published census volumes for 1860. Because county data is available for 1860, it is possible to confine the analysis to cotton producing regions. Two sets of regions are utilized in the following analysis. The first of these consists of the 409 "cotton counties" described in Chapter II.[1] This region is divided into two parts--Old South and New South. The Old South region consists of the Virginia, North Carolina, South Carolina, Georgia and Florida cotton counties. The New South region consists of the "cotton counties" in Tennessee, Alabama, Mississippi, Louisiana, Arkansas and Texas. The second set of regions consists of Gray's seventy-four major cotton producing counties described in Chapter III.[2] These counties are classified into three groups--alluvial regions, other New South regions, and Old South regions--according to soil and climatic conditions.

---

[1]See Chapter II, p. 25.

[2]See Chapter III, pages 51 and 52.

In Table 10, the size distributions of farms in the "cotton counties" of the Old and New South are contrasted. As the chi-square statistic indicates, the distributions of farm sizes are significantly different at the .0005 level due to the fact that smaller farms were relatively more numerous in the New South than in the Old South. In fact, 45.7 percent of the New South farms contained less than fifty improved acres as compared to 28.3 percent in the Old South.

TABLE 10

SIZE DISTRIBUTION OF FARMS BY REGION--1860
OLD SOUTH VS NEW SOUTH

Size of Farm* (Improved Acreage)		Old South		New South	
3-9	acres	1,291	(1.4%)	4,690	(2.7%)
10-19	acres	3,849	(4.2%)	18,012	(10.5%)
20-49	acres	20,735	(22.7%)	55,812	(32.5%)
50-99	acres	22,682	(24.8%)	40,592	(23.6%)
100-499	acres	36,461	(39.8%)	45,154	(26.3%)
500-999	acres	4,862	(5.3%)	6,016	(3.5%)
1000 acres & up		1,640	(1.8%)	1,695	(1.0%)
	Total	91,521	(100.0%)	171,971	(100.0%)
				$X^2(df=6) > 24.12$ $p < .0005$	

*Farms of less than three acres are not included in the published census tables.

Source: U.S. Bureau of the Census, Eighth Census of the United States: 1860, Agriculture.

Gray's three major cotton-producing regions--Old South, Alluvial, and New South--are contrasted in Table 11. Again, the farm size distribution in each region is significantly different from those in each of

TABLE 11

SIZE DISTRIBUTION OF FARMS BY REGION--1860
OLD SOUTH VS ALLUVIAL VS NEW SOUTH

Size of Farm* (Improved Acreage)	Old South		Alluvial		New South	
3-9 acres	163	(0.8%)	248	(2.1%)	372	(2.6%)
10-19 acres	522	(2.4%)	996	(8.5%)	1,061	(7.4%)
20-49 acres	3,315	(15.5%)	2,232	(19.1%)	4,070	(28.3%)
50-99 acres	4,635	(21.7%)	1,783	(15.2%)	3,230	(22.4%)
100-499 acres	10,247	(48.0%)	4,405	(37.7%)	4,548	(31.5%)
500-999 acres	1,790	(8.4%)	1,502	(12.8%)	801	(5.6%)
1000 acres & up	668	(3.1%)	527	(4.5%)	326	(2.3%)
Total	21,340	(100.0%)	11,693	(100.0%)	14,418	(100.0%)

Old South vs Alluvial: $X^2(df=6) > 24.12$, $p < .0005$

Alluvial vs New South: $X^2(df=6) > 24.12$, $p < .0005$

Old South vs New South: $X^2(df=6) > 24.12$, $p < .0005$

*Farms of less than three acres are not included in the published census tables.

Source: U.S. Bureau of the Census, _Eighth Census of the United States: 1860, Agriculture_.

the other two regions at a .0005 level of significance. In the newer regions (Alluvial and New South), small farms were relatively more numerous than in the Old South. This tendency toward smaller farms was most pronounced in the New South region where 38.3 percent of the farms contained less than fifty improved acres. In the Alluvial region, farms of less than fifty improved acres were more common than in the Old South and farms of five hundred or more improved acres were more common than in either the Old or New South regions. Thus, although the Alluvial region could be characterized as a region of large plantations, it exceeded the Old South in percentage of smaller farms.

In Table 12, the size distributions of slaveholdings in the New and Old South "cotton counties" are compared and, in Table 13, the distributions in Gray's three major cotton-producing regions are presented. Before examining these tables, two cautions should be raised. First, the number of nonslaveholders is not available in the published census volumes. Thus, a tremendous group of farmers is completely omitted. (As indicated in Chapter I, this omission is hardly insignificant!) Second, the <u>published</u> distributions include both <u>urban</u> and rural slaveholdings. Thus, it is impossible to determine whether differences between regions result from regional differences in wealth, regional differences in agricultural methods, or merely regional differences in urbanization.

Examination of Table 12 indicates a slight tendency toward smaller slaveholdings in the New South than in the Old South. Slaveholdings of five slaves or less constituted 49.0 percent of New South slaveholdings and only 45.8 percent of Old South slaveholdings in 1860. In Table 13, the size distributions of slaveholdings in Gray's major cotton producing

TABLE 12

SIZE DISTRIBUTION OF SLAVEHOLDINGS BY REGION--1860
OLD SOUTH VS NEW SOUTH

Size Distribution of Slaveholdings		Old South		New South	
1	slave	11,936	(15.2%)	19,936	(16.7%)
2	slaves	7,892	(10.0%)	12,952	(10.8%)
3	slaves	6,231	(7.9%)	10,078	(8.4%)
4	slaves	5,365	(6.8%)	8,458	(7.1%)
5	slaves	4,657	(5.9%)	7,192	(6.0%)
6	slaves	4,114	(5.2%)	6,103	(5.1%)
7	slaves	3,556	(4.5%)	5,041	(4.2%)
8	slaves	3,215	(4.1%)	4,611	(3.9%)
9	slaves	2,770	(3.5%)	3,881	(3.3%)
10-14	slaves	9,597	(12.2%)	13,427	(11.2%)
15-19	slaves	5,652	(7.2%)	7,625	(6.4%)
20-29	slaves	5,904	(7.5%)	8,141	(6.8%)
30-39	slaves	2,954	(3.7%)	4,169	(3.5%)
40-49	slaves	1,629	(2.1%)	2,406	(2.0%)
50-69	slaves	1,689	(2.1%)	2,521	(2.1%)
70-99	slaves	941	(1.2%)	1,707	(1.4%)
100-199	slaves	588	(0.7%)	1,014	(0.8%)
200-299	slaves	68	(0.1%)	100	(0.1%)
300-499	slaves	20	---	27	---
500-999	slaves	5	---	4	---
1000 slaves & over		---	---	---	---
		78,783	(100.0%)	119,393	(100.0%)

Source: U.S. Bureau of the Census, *Eighth Census of the United States: 1860, Agriculture*.

TABLE 13

SIZE DISTRIBUTION OF SLAVEHOLDERS BY REGION--1860
OLD SOUTH VS ALLUVIAL VS NEW SOUTH

Size Distribution of Slaveholdings		Old South		Alluvial		New South	
1	slave	2,724	(12.7%)	1,877	(13.2%)	1,793	(14.8%)
2	slaves	1,897	(8.9%)	1,276	(9.0%)	1,223	(10.1%)
3	slaves	1,582	(7.4%)	959	(6.7%)	919	(7.6%)
4	slaves	1,348	(6.3%)	820	(5.8%)	832	(6.9%)
5	slaves	1,190	(5.6%)	729	(5.1%)	665	(5.5%)
6	slaves	1,123	(5.2%)	602	(4.2%)	600	(4.9%)
7	slaves	935	(4.4%)	509	(3.6%)	493	(4.1%)
8	slaves	890	(4.2%)	480	(3.4%)	470	(3.9%)
9	slaves	755	(3.5%)	369	(2.6%)	386	(3.2%)
10-14	slaves	2,741	(12.8%)	1,449	(10.2%)	1,430	(11.8%)
15-19	slaves	1,653	(7.7%)	924	(6.5%)	849	(7.0%)
20-29	slaves	1,852	(8.7%)	1,281	(9.0%)	882	(7.3%)
30-39	slaves	1,020	(4.8%)	746	(5.2%)	521	(4.3%)
40-49	slaves	557	(2.6%)	514	(3.6%)	331	(2.7%)
50-69	slaves	591	(2.8%)	706	(5.0%)	309	(2.5%)
70-99	slaves	353	(1.6%)	537	(3.8%)	243	(2.0%)
100-199	slaves	177	(0.8%)	377	(2.6%)	160	(1.3%)
200-299	slaves	15	(0.1%)	54	(0.4%)	21	(0.2%)
300-499	slaves	5	---	13	(0.1%)	7	(0.1%)
500-999	slaves	---	---	3	---	---	---
1000 slaves & over		---	---	---	---	---	---
Total		21,408	(100.0%)	14,255	(100.0%)	12,134	(100.0%)

Source: U.S. Bureau of the Census, <u>Eighth Census of the United States: 1860, Agriculture</u>.

regions reveal this same tendency toward smaller slaveholdings in the newer regions (Alluvial and New South). Consistent with the results regarding size distribution of farms, the Alluvial region not only exhibits a tendency toward relatively more slaveholdings of three slaves or less but also a greater percentage of slaveholdings of twenty slaves or more. In the Alluvial region, 29.7 percent of all slaveholdings included twenty or more slaves. In the Old and New South regions the percentages were 21.4 and 20.4 percent respectively. Again, the Alluvial region could be characterized as a region of large slaveholdings, yet, it also exceeded the Old South in the percentages of slaveholdings in the one- and two-slave classes.

From the above discussion we can conclude that smaller farms and smaller slaveholdings were relatively more numerous in the Southwest than in the Southeast in 1860. However, in the Alluvial regions, smaller farms and smaller slaveholdings were more common than in the Old South but the percentages of farms and slaveholdings in the largest classes exceeded those in either the Old South or New South regions. Unless one is willing to argue that the large planters who moved to the newer regions operated smaller farms than they had operated in the Old South, the first part of our hypothesis appears questionable. (The possibility that planters sold large farms in the Old South and moved west with the cash proceeds will be discussed, and rejected, in Chapter V.) If large planters led the migration, one would hardly expect to find a greater percentage of small farmers in the newer regions than in the old.

The fact that very large farms and very large slaveholdings were more common in the Alluvial than in the New or Old South may indicate the source of our hypothesis. If historians have tended to concentrate their efforts on the Alluvial regions, the conclusion that large planters led the migration appears reasonable.

Tests of the Hypothesis--1850 Data

Testing the first part of our hypothesis using data from the published census volumes for 1850 is a very brief task due to the paucity of useful data. For 1850, only size distributions of slaves are available and these distributions are available only for entire states. Thus, it is impossible to confine the analysis to cotton-producing regions or to consider regional differences within states. Additionally, the data include both urban and rural slaveholdings and ignore the nonslaveholders completely.

In Table 14, the major cotton-producing states have been divided into Old South and New South states using the Alabama-Georgia boundary to effect the division. States east of this boundary have been considered Old South; to the west, New South. Slaveholdings of only one slave were relatively more numerous in the New South than in the Old South. In fact, over twenty percent of the New South slaveholdings consisted of only one slave as contrasted with under thirteen percent in the Old South. Further, slaveholdings of two hundred or more slaves were much more common in the Old South than in the New South in 1850.

Once again, the limited data available in the published census suggest that the smaller operators were relatively more numerous in the New South than in the Old South. Thus, the 1850 data fail to support the hypothesis that the larger planters led the migration to the Southwest.

TABLE 14

SIZE DISTRIBUTION OF SLAVEHOLDINGS BY REGION--1850
OLD SOUTH VS NEW SOUTH

Size Distribution of Slaveholdings		Old South*		New South**	
1	slave	11,949	(12.5%)	24,575	(20.4%)
2-4	slaves	28,539	(29.8%)	35,210	(29.2%)
5-9	slaves	22,900	(23.9%)	27,306	(22.6%)
10-19	slaves	17,931	(18.7%)	18,495	(15.3%)
20-49	slaves	11,433	(11.9%)	11,220	(9.3%)
50-99	slaves	2,343	(2.4%)	3,062	(2.5%)
100-199	slaves	634	(0.7%)	726	(0.6%)
200-299	slaves	103	(0.1%)	75	(0.1%)
300-499	slaves	37	---	17	---
500-999	slaves	4	---	5	---
1000 slaves & up		2	---	---	---
	Total	95,875	(100.0%)	120,691	(100.0%)

*Florida, Georgia, North Carolina, and South Carolina.

**Alabama, Arkansas, Louisiana, Mississippi, Tennessee, and Texas.

Source: Calculated from U.S. Bureau of the Census, Eighth Census of the United States: 1860, Agriculture.

Regarding Displacement of the Small Farmer

A very limited test of the second part of our hypothesis--that the small farmers were pushed out of the newer regions by larger planters who were able to buy the better land from them--may be effected using data from the published census volumes.

In Table 15, the size distributions of slaveholdings in the Old South in 1850 and 1860 are contrasted as are the size distributions in the New South for the two years. In each region, there was little change in the size distribution of slaveholdings between 1850 and 1860. The percentage of slaveholdings of one slave did increase from 12.5 percent to a level of 16.5 percent in the Old South but this increase is fully offset by a decline in the percentage of slaveholdings in the two to four slave category. In the New South, the percentage of slaveholdings in both of these categories declined. Thus, this very limited evidence would suggest the possibility that small slaveholders were being bought out by larger planters in the New South. Of course, it is also possible that the small slaveholders prospered during the decade and purchased additional slaves.

TABLE 15

SIZE DISTRIBUTION OF SLAVEHOLDERS
1850 VS 1860

Size Distribution by Slaveholdings		Old South* 1850	Old South* 1860	New South** 1850	New South** 1860
1	slave	(12.5%)	(16.5%)	(20.4%)	(18.6%)
2-4	slaves	(29.8%)	(26.1%)	(29.2%)	(28.1%)
5-9	slaves	(23.9%)	(23.6%)	(22.6%)	(22.8%)
10-19	slaves	(18.7%)	(18.4%)	(15.3%)	(16.2%)
20-49	slaves	(11.9%)	(11.7%)	(9.3%)	(10.5%)
50-99	slaves	(2.4%)	(2.9%)	(2.5%)	(3.0%)
100-199	slaves	(0.7%)	(0.7%)	(0.6%)	(0.8%)
200-299	slaves	(0.1%)	(0.1%)	(0.1%)	(0.1%)
300-499	slaves	---	---	---	---
500-999	slaves	---	---	---	---
1000 slaves & over		---	---	---	---
	Total	(100.0%)	(100.0%)	(100.0%)	(100.0%)

*Florida, Georgia, North Carolina and South Carolina.

**Alabama, Arkansas, Louisiana, Mississippi, Tennessee and Texas.

Source: Calculated from U.S. Bureau of the Census, *Eighth Census of the United States: 1860, Agriculture*.

Unfortunately, it is impossible to reach any conclusions regarding the position of small farmers _vis a vis_ large planters using the data in Table 15. For one thing, the data are state wide data combining everything from alluvial river bottoms to frontier grasslands into one category. Second, the data include both urban and rural slaveholdings making any conclusions tenuous, especially where slaveholdings of one or two slaves are involved.

An indication of the difficulties which can be brought about by urban-rural differences may be obtained by considering the growth of cities and towns in the New and Old South between 1850 and 1860. In the Old South (Florida, Georgia, North Carolina, and South Carolina), total population increased 14.3 percent between 1850 and 1860. During the same period, cities and towns with populations of over 1,000 persons in 1860 increased their populations by 22.1 percent. (Actually these data apply only to cities and towns for which population data were available for both 1850 and 1860. If one assumes that many towns were omitted from the 1850 census because they were too small to be considered important, the percentage change would be somewhat greater. An upper limit for this latter percentage change is 60.9 percent. This bound is calculated by taking population data for all towns above 1,000 population in 1860 and comparing this total to the total for any of these same towns which happen to be included in the 1850 census.) In the New South (Alabama, Arkansas, Louisiana, Mississippi, Tennessee, and Texas), total population increased 38.9 percent between 1850 and 1860. Population of cities and towns with populations of over 1,000 persons in 1860 increased 51.5 percent. (The upper limit for this increase is 77.0 percent.)

Thus, Old South cities grew 1.5 times the increase in Old South population while New South cities grew at 1.3 times the increase in New South population. (The upper limits for these multiples are: Old South - 4.3; New South - 2.0.) The fact that cities and towns were growing faster, relative to rural growth, in the Old South than in the New South may mean that urban slaveholders were becoming a greater proportion of total slaveholders in the Old South and a lesser proportion of total slaveholders in the New South. If this happened to be the case, one would expect smaller slaveholdings to increase relatively in the Old South and decrease in the New South.

The above discussion is not intended to fully explain the results in Table 15 but, rather, it is intended to illustrate the difficulties involved with using statewide data. Unfortunately, the only way to avoid these difficulties is to sift the manuscript census returns.

THE ROLE OF THE YEOMAN FARMER

In addition to testing the two-part hypothesis concerning the participation of smaller farmers in the southwestward migration, this study attempts to ascertain the contribution of the small farmer to the antebellum southern economy. In this regard, the published census data is of little value. Crop production, by county, were included in the published volumes for both 1850 and 1860. Additionally, a great variety of data on farm sizes and wealth holdings were also published. However, the Census Bureau made no attempt to report the crop outputs or wealth holdings by size of farm or size of slaveholding. That is, no attempt was made to determine the contribution of any economic group to the total output of any one county or the share of the wealth held by any group in any one county.

This dearth of data makes it impossible even to begin to determine the role of the smaller farmer in the antebellum southern economy using published information. Thus, this analysis is postponed until Chapter VIII.

DATA CONTRIBUTION OF THIS STUDY

From the above paragraphs it should be apparent that an analysis of the role of the smaller farmer in southwestward migration and in the antebellum southern economy requires going beyond the data which is readily available in the published census volumes. Thus, it is necessary to go to the very extensive record left by all farmers, small or large, in the manuscript censuses of agriculture and population.

By selecting large, extensive, and representative samples from the manuscript censuses of 1850 and 1860, new data has been made available to analyze the role of the smaller farmer in westward migration and in the antebellum southern economy. The analysis in the following chapters utilizes at least four sets of data which were not previously available.

First, size distributions of farms for 1850 are available only for a limited number of states. Thus, it is impossible to compare regions other than a few entire states in 1850 or to consider any subregion such as primary cotton-growing regions without turning to the manuscript census data. (In 1860, size distributions of farms, by county, were included in the published census.)

Second, size distributions of slaveholdings for 1850 are available only on a state level. It is impossible to compare regions other than entire states without using the manuscript census. (In 1860, size distributions of slaveholdings were published for each county.)

Additionally, the distributions in the published census volumes include both urban and rural slaveholdings. The present study requires that rural slaveholdings be isolated and makes size distributions of rural slaveholdings available for a variety of regions. These distributions are now available for future research.

Third, no data exist regarding nonslaveholders. (This explains why the very large percentage of nonslaveholders in the Southwest came as such a surprise!) Granted, it would be possible to infer approximate numbers of nonslaveholders from the published data on slaveholders, numbers of farms, population, etc. but such inferences would require numerous assumptions, e.g., the ratio of urban to rural slaveholdings. The present study spotlights the large number of nonslaveholding farmers.

Last, and perhaps most significant, the manuscript census samples employed in this study make it possible to evaluate the contributions of various groups of farmers to the antebellum southern economy. In Chapter VIII, the percentages of various crops produced by small landowners, nonslaveholders, and small slaveholders are presented and evaluated. To the writer's knowledge, no other source of crop output data by size of farm or size of slaveholding exists.

Thus, the present study is an analysis of data heretofore unavailable and the presentation of that data for the perusal of other scholars.

CHAPTER V

TESTING THE HYPOTHESIS--1860

The preceding chapters describe the 1850 and 1860 manuscript census samples and hopefully, have convinced the reader that the samples are sufficiently representative to be used to test the two-part hypothesis specified in Chapter I. In this chapter we will begin to test the first part of the hypothesis--that the migration into the new land in the Southwest was one of large planters leading, accompanying, or *immediately* following smaller farmers. We will examine the data from the 1860 manuscript census and compare the size of farm and slaveholdings in the newer regions with those of the older regions. Also, we will compare the role of the smaller farmers of the newer regions with that of their counterparts in the old. But, we're getting ahead of ourselves. First we should define, for the purposes of this paper, what we mean by the older and newer regions. Also, we need to take a second close look at each of our samples to determine whether they are representative of each of the regions we wish to examine. If not, we shall have to keep the regional biases of the sample in mind as we proceed with our analysis.

DEFINITION OF REGIONS

The hypotheses specified in Chapter I will be tested by examining the distribution of farm sizes and slaveholdings in the newer (Southwestern) and older (Southeastern) regions in the census years 1850 and 1860. The two different sets of regions used in testing the hypotheses are described in the following paragraphs. The first set of regions is used only to examine the 1860 manuscript census data and make use of all 5,229 farms included in that sample. The second breakdown covers a much more limited region and thus utilizes only a 964 farm subsample of the 1860 sample. The more limited 1850 sample is selected from this latter region only.

The first regional breakdown which is employed involves splitting the 5,229 farms of the 1860 sample into two parts using the Alabama-Georgia state boundary as the dividing line. (This simple division is used to facilitate handling the data. However, it will be argued below that this is a reasonable as well as convenient division.) The states to the east of this boundary, Florida, Georgia, North Carolina, South Carolina, and Virginia, make up the Old South region. The New South region consists of Alabama, Arkansas, Louisiana, Mississippi, Tennessee, and Texas.

The reader may well question whether the Alabama-Georgia boundary truly divides the South into a newer and older region. If we take the increase in the number of farms three acres and larger between 1850 and 1860 as indicative of the "newness" of a region, this division appears to be a fairly reasonable one. The "cotton counties" in the Old South region, as defined above, contained 90,968 farms of three acres and larger in 1850 and 91,521 in 1860, an increase of 0.6 percent. The New South region contained 128,841 farms three acres and larger in 1850. This number grew to 171,976 farms in 1860, an increase of 33.5 percent. In only one Old South state, North Carolina, was the increase in the number of cotton county farms greater than 10 percent and two of the Old South states actually showed a reduction in the number of farms between the two census years. (See Table 16.) On the other hand, in four of the

New South states the number of farms increased by more than 20 percent. Tennessee was the only New South state in which the number of farms decreased between the two census years.[1] Thus, it is argued that this New South-Old South division is satisfactory for the purposes of this paper.

TABLE 16

NUMBER OF FARMS THREE ACRES AND LARGER IN COUNTIES PRODUCING AT LEAST 1,000 BALES OF COTTON IN 1859

Region	1850	1860	% Increase (Decrease)
Old South			
Florida	3,008	3,243	7.8%
Georgia	39,440	38,678	(1.9)
North Carolina	18,963	21,201	11.8
South Carolina	27,457	26,228	(4.3)
Virginia	2,100	2,171	3.4
Total	90,968	91,521	.6
New South			
Alabama	40,344	49,357	22.3
Arkansas	11,219	21,781	94.1
Louisiana	10,521	13,979	32.9
Mississippi	33,021	36,235	9.7
Tennessee	23,580	23,536	(.2)
Texas	10,156	27,088	166.7
Total	128,841	171,976	33.5

Source: 1850--U.S. Bureau of the Census, *Seventh Census of the United States: 1850, Compendium.*
1860--U.S. Bureau of the Census, *Eighth Census of the United States: 1860, Agriculture.*

[1]The 465 sample farms selected from Tennessee are included in the New South due to their proximity to the New South farms selected from other states. The Tennessee "cotton counties" are located in the southwestern and southcentral portions of the state, adjacent to the Mississippi and Alabama borders.

The other regional breakdown employed in this paper is that used by Gray in describing his seventy-four major cotton producing counties. According to Gray, the counties which lie in South Carolina and Georgia are Old South counties; those which lie along the river valleys of Louisiana, Mississippi, Arkansas and northern Alabama are alluvial counties; those in eastern Texas and in the Alabama "Black Belt" are New South counties. A comparison of the number of farms three acres and larger in these regions in 1850 and 1860 is given in Table 17. In the Old South counties, the number of farms _decreased_ by 14.5 percent between 1850 and 1860. In the Alluvial and New South counties the number of farms _increased_ by 29.4 and 20.1 percent respectively. Again, it seems reasonable to assume that Gray's Old South counties are located in a region which was settled much earlier than were the Alluvial and New South counties and, consequently, comparison between these three regions should prove useful in testing our hypothesis.

TABLE 17

NUMBER OF FARMS THREE ACRES AND LARGER IN GRAY'S SEVENTY-FOUR MAJOR COTTON PRODUCING COUNTIES

Region	1850	1860	% Increase (Decrease)
Old South	24,973	21,340	(14.5)
Alluvial	9,362	11,693	24.9
New South	12,005	14,418	20.1

Source: 1850--U.S. Bureau of the Census, _Seventh Census of the United States: 1850, Compendium._
1860--U.S. Bureau of the Census, _Eighth Census of the United States: 1860, Agriculture._

INTERREGIONAL COMPARISONS

The remainder of this chapter is devoted to a comparison of the role of the small farmer in the newer areas with his role in the older areas. First, the two regions formed by dividing the South at the Alabama-Georgia border are considered. In the final pages of this chapter the data selected from the Old South, Alluvial, and New South regions, as described by Gray, are considered. The reader should be reminded again that all of the data have been selected from counties which produced at least one thousand bales of cotton in 1859. Thus, any conclusions apply only to the cotton producing regions.

New South vs Old South

The first comparison of the role of the small farmer in the New and the Old South utilizes the entire 5,229 farm 1860 sample dividing it at the Alabama-Georgia boundary. The New South sample, as defined, contains 3,387 farms; the Old South, 1,841. (One farm was accidentally omitted.)

Before evaluating the role of the small farmer in each of the regions, we should once again test the data in an attempt to estimate whether each sample is representative of the region from which it is selected. The first test of the regional samples is a comparison of the sample mean with the universe mean using the t-statistic which we employed earlier in the paper. The results of this test for fifteen items of property and crop data are provided in Table 18. In each region, only one of the fifteen sample means is significantly different from the universe mean if a five percent level of significance ($|t| > 1.96$) is employed. In the Old South region corn production per sample farm was significantly lower than corn production on the universe farms. Unfortunately, in the New

TABLE 18

TESTS OF THE 1860 SAMPLE BY REGION
OLD SOUTH VS. NEW SOUTH
(Acreage, Value of Farm, Crop Outputs)

	Old South Region	New South Region
Number of farms in sample	1841	3387
Values of t - statistics		
Improved acres	-.93	-.85
Unimproved acres	-1.13	-.48
Value of farm	-1.52	1.07
Value of farm machinery	-1.88	-.07
Swine	-1.20	-1.62
Value of livestock	-1.95	-.69
Bushels of corn	-2.50	-.29
Bales of cotton	-.81	-2.21
Bushels of peas & beans	-.55	.73
Bushels of Irish potatoes	.70	-1.65
Bushels of sweet potatoes	-.49	.46
1000# of cane sugar	1.00	.21
Gallons of molasses	.35	.44
Value of home manufactures	-.52	-.70
Value of animals slaughtered	.10	.08

Source: Calculated from 1860 manuscript census sample and U.S. Bureau of the Census, Eighth Census of the United States: 1860, Agriculture.

South cotton production per sample farm was significantly lower than the universe mean. In the Old South, eleven of the fifteen sample crop outputs were smaller than the universe outputs. In the New South, nine of the fifteen were lower than their universe counterparts. This would imply that crop outputs in the Old South sample are biased toward the low side. (The probability of getting eleven negative and four positive values of the t-statistic is approximately .06). In the New South a sign test is not significant.

The second test of the regional samples compares the observed and expected distribution of the farm sizes. The distribution of farm sizes in each sample is not significantly different from the expected distribution if a five percent level of significance is employed. (See Table 19.) The size distribution of farms in the New South sample is practically identical to the size distribution in the New South universe. In the Old South sample, there are an excessive number of farms of 1000 acres and larger. However, this excess is not sufficient to significantly bias the sample.

The third test of the regional samples involves a comparison of the size distribution of slaveholdings on the sampled farms with that of the universe. In each regional sample there is a deficit of slaveholdings in the small categories. (See Table 20.) As indicated in earlier chapters, this result is expected due to the fact that the sample is selected entirely from rural regions.

The above tests provide insufficient evidence to reject the New South or the Old South sample. Thus, we accept each of the samples as being representative of their respective universes and move on to a comparison of farm sizes and slaveholdings in the New South and Old South regions.

TABLE 19

TESTS OF THE 1860 SAMPLE BY REGION
OLD SOUTH VS NEW SOUTH
(Size Distribution of Farms)

Size of Farm	Old South Region Observed	Old South Region Expected	New South Region Observed	New South Region Expected
3-9 acres	23	25.7	85	90.8
10-19 acres	63	76.6	340	348.9
20-49 acres	422	412.8	1058	1081.1
50-99 acres	475	451.6	805	786.2
100-499 acres	710	725.9	897	874.6
500-999 acres	84	96.8	112	116.5
1000 acres & up	45	32.6	34	32.8
Total*	1822	1822.0	3331	3330.9
	$x^2(df=6) = 10.87$ $p = .09$		$x^2(df=6) = 2.32$ $p = .89$	

*Farms of less than three acres are not included in the table. The number of sample farms having less than three acres of improved land is:

Old South 19
New South 56

Source: Calculated from the 1860 manuscript census sample and U.S. Bureau of the Census, Eighth Census of the United States: 1860, Agriculture.

TABLE 20

TESTS OF THE 1860 SAMPLE BY REGIONS
OLD SOUTH VS. NEW SOUTH
(Size Distribution of Slaveholdings)

Size of slaveholding		Old South Region Observed	Old South Region Expected	New South Region Observed	New South Region Expected
1	slave	119	163.6	175	251.6
2	slaves	88	108.2	131	163.5
3	slaves	63	85.4	97	127.2
4	slaves	66	73.5	85	106.8
5	slaves	71	63.8	74	90.8
6	slaves	42	56.4	71	77.0
7	slaves	52	48.7	50	63.6
8	slaves	61	44.1	63	58.2
9	slaves	32	38.0	52	49.0
10-14	slaves	143	131.6	192	169.5
15-19	slaves	99	77.5	138	96.2
20-29	slaves	92	80.9	151	102.8
30-39	slaves	58	40.5	95	52.6
40-49	slaves	35	22.3	44	30.4
50-69	slaves	27	23.2	40	31.8
70-99	slaves	17	12.9	25	21.5
100-199	slaves	14	8.1	23	12.8
200-299	slaves	1	.9	--	1.3
300-499	slaves	--	.3	1	.3
500-999	slaves	--	.1	--	.1
1000 slaves & larger		--	--	--	--
	Total	1080	1080.0	1507	1507.0
		$x^2(df=17) > 42.84$ $p < .0005$		$x^2(df=17) > 42.84$ $p < .0005$	
Farms without slaves		761		1880	

Source: Calculated from 1860 manuscript census sample and U.S. Bureau of the Census, Eighth Census of the United States: 1860, Agriculture.

The size distributions of sample farms in the Old and New South in 1860 are compared in Table 21. Smaller farms were relatively more common in the New South region. In this region, 14.2 percent of the farms contained less than twenty acres, 45.4 percent less than fifty acres, and 69.2 percent less than one hundred acres. In the Old South, only 5.6 percent of the sample farms contained less than twenty acres, 28.5 less than fifty and 54.3 percent in less than one hundred. The distributions of farm sizes in the two regions are significantly different at the .01 level.

TABLE 21

SIZE DISTRIBUTION OF FARMS (IMPROVED ACREAGE)
OLD SOUTH VS NEW SOUTH--1860

Farm Size	Old South		New South	
Less than 3 acres	19	(1.0%)	56	(1.7%)
3-9 acres	23	(1.2%)	85	(2.5%)
10-19 acres	63	(3.4%)	340	(10.0%)
20-49 acres	422	(22.9%)	1058	(31.2%)
50-99 acres	475	(25.8%)	805	(23.8%)
100-199 acres	406	(22.0%)	509	(15.0%)
200-299 acres	161	(8.7%)	198	(5.8%)
300-499 acres	143	(7.7%)	190	(5.6%)
500-999 acres	84	(4.5%)	112	(3.3%)
1000 acres & larger	45	(2.4%)	34	(1.0%)
Total	1841		3387	

$x^2(df=9) > 23.6$
$p < .005$

Source: Calculated from 1860 Manuscript Census Sample.

As can be seen in Table 22, nonslaveholders were relatively more numerous in the New South than in the Old South. Only 41 percent of the Old South farms were slaveless as compared to over 55 percent of the New South farms. Again the size distributions of slaveholdings in the two regions are significantly different at the .01 level. However, if non-slaveholdings are eliminated from the size distribution of slaveholdings, the resulting distributions in the two regions are practically identical. (See Table 23.) Slaveholdings of 1-3 slaves appear to have been slightly more common in the New South than in the Old South (26.7 percent vs. 24.9 percent). However, the tendency toward smaller slaveholdings is not statistically significant.

Thus, we find that small landowners were a greater percentage of farmers in the New South than in the Old South. Further, nonslaveholdings were predominant in the New South. Finally, small slaveowners show some tendency to be more common in the New South than in the Old South but this tendency is not statistically significant. At this point, the hypothesis that large planters led the migration into the Southwest appears somewhat shaky.

Next, we might take a closer look at each of the samples in an attempt to determine what contribution the small farms made to their respective regions. Crop outputs, acreage, and property data by acreage class expressed as a percentage of regional totals are given in Table 24. In every category except sugar and molasses production, the share of the total New South property owned or output produced by farmers with less than twenty improved acres exceeded the share of total Old South property owned or crops produced by Old South farmers who operated farms of less than twenty acres. The same conclusion obtains if farmers with less than

TABLE 22

DISTRIBUTION OF SLAVEHOLDINGS
OLD SOUTH VS NEW SOUTH - 1860

Slaveholdings		Old South		New South	
0	slaves	761	(41.3%)	1880	(55.5%)
1	slave	119	(6.4%)	175	(5.2%)
2	slaves	88	(4.7%)	131	(3.9%)
3	slaves	63	(3.4%)	97	(2.9%)
4	slaves	66	(3.5%)	85	(2.5%)
5	slaves	71	(3.8%)	74	(2.2%)
6	slaves	42	(2.2%)	71	(2.1%)
7	slaves	52	(2.8%)	50	(1.5%)
8	slaves	61	(3.3%)	63	(1.9%)
9	slaves	32	(1.7%)	52	(1.5%)
10-14	slaves	143	(7.7%)	192	(5.7%)
15-19	slaves	99	(5.3%)	138	(4.1%)
20-29	slaves	92	(4.9%)	151	(4.5%)
30-39	slaves	58	(3.1%)	95	(2.8%)
40-49	slaves	35	(1.9%)	44	(1.3%)
50-69	slaves	27	(1.4%)	40	(1.2%)
70-99	slaves	17	(.9%)	25	(.7%)
100-199	slaves	14	(.7%)	23	(.6%)
200-299	slaves	1	(.0%)	0	(.0%)
300-499	slaves	0	(.0%)	1	(.0%)
500-999	slaves	0	(.0%)	0	(.0%)
1000 slaves & up		0	(.0%)	0	(.0%)
	Total	1841		3387	

$x^2 (df=17) > 35.7$
$p < .005$

Source: Calculated from 1860 Manuscript Census Sample.

TABLE 23

DISTRIBUTION OF SLAVEHOLDINGS (EXCLUDING NONSLAVEHOLDERS)
OLD SOUTH VS NEW SOUTH - 1860

Slaveholdings		Old South		New South	
1	slave	119	(11.0%)	175	(11.6%)
2	slaves	88	(8.1%)	131	(8.7%)
3	slaves	63	(5.8%)	97	(6.4%)
4	slaves	66	(6.1%)	85	(5.6%)
5	slaves	71	(6.6%)	74	(4.9%)
6	slaves	42	(3.9%)	71	(4.7%)
7	slaves	52	(4.8%)	50	(3.3%)
8	slaves	61	(5.6%)	63	(4.2%)
9	slaves	32	(3.0%)	52	(3.5%)
10-14	slaves	143	(13.2%)	192	(12.7%)
15-19	slaves	99	(9.2%)	138	(9.2%)
20-29	slaves	92	(8.5%)	151	(10.0%)
30-39	slaves	58	(5.4%)	95	(6.3%)
40-49	slaves	35	(3.2%)	44	(2.9%)
50-69	slaves	27	(2.5%)	40	(2.7%)
70-99	slaves	17	(1.6%)	25	(1.7%)
100-199	slaves	14	(1.3%)	23	(1.5%)
200-299	slaves	1	(.1%)	0	(.0%)
300-499	slaves	0	(.0%)	1	(.1%)
500-999	slaves	0	(.0%)	0	(.0%)
1000 slaves & up		0	(.0%)	0	(.0%)
	Total	1080		1507	

x^2(df=16) = 14.9
p = .53

Source: Calculated from 1860 Manuscript Census Sample.

TABLE 24

CROP OUTPUT, WEALTH, ACREAGE, AND POPULATION BY ACREAGE CLASSES
AS A PERCENT OF REGIONAL TOTALS
OLD SOUTH VS NEW SOUTH-1860

		Farm Sizes (Improved Acreage)							
Item	Region	Less than 20	20-49	50-99	100-199	200-299	300-499	500-999	1000 & Larger
No. of Farms	OS	5.6%	22.9%	25.8%	22.1%	8.7%	7.8%	4.6%	2.4%
	NS	14.2	31.2	23.8	15.0	5.8	5.6	3.3	1.0
Val. of Pers. Prop.	OS	.6	3.5	7.0	17.5	13.5	20.6	18.3	19.0
	NS	1.6	5.0	9.0	17.1	14.0	20.6	20.9	11.8
Imp. Acreage	OS	.3	4.6	10.4	17.5	12.1	16.8	17.8	20.5
	NS	1.3	8.6	13.7	16.8	11.5	17.7	18.2	12.0
Unimp. Acreage	OS	1.1	8.8	15.5	20.2	14.4	16.9	13.0	10.1
	NS	7.6	14.3	18.2	17.6	9.5	12.4	11.4	9.0
Val. of Farm	OS	.6	5.2	12.0	18.4	12.7	17.1	17.5	16.4
	NS	1.8	6.4	9.1	14.6	10.0	16.6	22.6	19.0
Horses	OS	2.1	13.1	20.2	24.9	10.6	12.2	9.5	7.4
	NS	7.5	20.5	19.9	16.8	8.8	15.0	8.0	3.5
Asses & Mules	OS	.6	4.1	8.9	15.4	12.5	18.4	19.7	20.4
	NS	7.5	5.3	9.3	15.2	12.5	18.4	19.6	12.3
Swine	OS	1.9	11.4	17.0	20.4	10.8	14.1	13.7	10.8
	NS	6.2	18.2	19.1	19.2	9.7	12.8	10.4	4.3
Val. of Livestock	OS	1.2	8.4	14.3	20.5	11.5	15.0	16.1	13.2
	NS	4.7	13.4	15.9	17.6	12.3	14.8	14.0	7.3

TABLE 24 (Continued)

Item	Region	Farm Sizes (Improved Acreage) Less than 20	20-49	50-99	100-199	200-299	300-499	500-999	1000 & Larger
Corn	OS	1.1%	7.6%	14.5%	20.4%	11.9%	16.0%	14.6%	13.9%
	NS	3.2	11.7	15.1	18.1	11.6	16.5	15.9	7.9
Tobacco	OS	3.0	5.0	6.5	28.8	28.1	27.9	.1	.6
	NS	6.9	27.7	28.0	25.8	6.6	4.7	.3	0
Cotton	OS	.4	3.9	8.2	14.2	10.5	15.4	21.5	25.9
	NS	1.0	5.3	8.3	14.2	12.4	21.4	22.2	15.3
Peas and Beans	OS	2.1	10.4	13.8	19.3	12.5	12.5	13.5	15.8
	NS	2.8	14.9	15.5	16.8	11.8	16.4	15.1	6.6
Irish Potatoes	OS	4.0	9.2	13.5	24.5	15.4	12.6	10.0	10.8
	NS	6.0	19.8	14.8	17.9	9.2	12.8	13.5	6.0
Sweet Potatoes	OS	2.3	11.8	15.8	21.1	12.8	11.8	12.0	12.3
	NS	4.5	15.8	17.0	18.1	10.2	14.6	13.2	6.6
Cane Sugar	OS	0	.8	2.5	1.3	6.4	2.5	85.6	.8
	NS	0	0	0	48.7	3.1	0	6.4	41.8
Molasses	OS	.2	10.7	20.5	8.8	33.6	11.4	8.6	6.1
	NS	.2	1.1	1.2	49.2	3.2	.8	10.8	33.5
Val. of Home Manu.	OS	3.2	15.6	23.0	21.3	9.6	13.1	6.4	7.8
	NS	8.8	25.9	24.2	16.0	8.9	7.6	6.5	2.1
Val. of An. Slau.	OS	1.8	9.6	15.7	28.7	11.9	12.1	10.6	9.6
	NS	5.6	15.5	18.6	17.2	10.3	14.4	12.4	5.8
Free Males	OS	4.5	21.6	27.4	22.8	8.3	8.3	4.6	2.5
	NS	11.4	30.9	26.7	16.3	5.6	5.0	3.3	.8

TABLE 24 (Continued)

Item	Region	Farm Sizes (Improved Acreage): Less than 20	20-49	50-99	100-199	200-299	300-499	500-999	1000 & Larger
Free Females	OS	5.1%	22.0%	27.5%	22.0%	8.4%	7.7%	4.3%	3.0%
	NS	12.4	32.5	25.9	15.1	5.2	5.0	3.1	.8
Slave Males	OS	.5	2.2	5.7	16.2	13.7	19.4	20.1	22.1
	NS	.6	2.4	7.0	16.6	14.7	21.6	22.9	14.2
Slave Females	OS	.3	2.4	6.2	17.4	14.4	18.0	21.2	20.0
	NS	.8	2.8	7.5	16.9	14.3	21.2	23.0	13.5
Total Population*	OS	2.2	9.9	14.4	19.0	11.8	14.5	14.3	13.9
	NS	5.8	15.9	16.0	16.3	10.3	13.9	13.9	7.9

*Total of four preceding items.

Source: Calculated from 1860 manuscript census sample.

fifty acres are compared. The differences in the contribution of small farms in the two regions are particularly striking if unimproved acreage, swine production, value of livestock owned, corn production, and tobacco production are considered. Also, the share of free population living on farms of less than twenty acres was two and one half times as great in the New South as in the Old South.

Crop outputs, acreage, and property data by slaveholding class are expressed as a percentage of regional totals in Table 25. The differences in the contributions of smaller slaveholdings in the two regions are not so striking as the differences in the relative contributions of the smaller farms. Nonslaveholders held a greater share of property and produce a greater percentage of crops in the New South than in the Old South in all categories except value of farm, and sugar and molasses production. However, if slaveholdings of less than five slaves (including nonslaveholders) are compared the differences between the two regions are much less striking. Slaveholdings of less than five slaves produced a greater share of swine, corn, sweet potatoes and home manufactures in the New South than did their counterparts in the Old. Additionally, the share of the New South personal estate, improved acreage, unimproved acreage, and livestock owned by these small slaveholders exceeded the share these wealth items held by small slaveholders in the Old South. In the New South, over seventy percent of the free population (rural) were associated with slaveholdings of less than five slaves while only sixty-one percent of the free population (rural) in the Old South lived on farms with less than five slaves. Slaveholders of less than five slaves owned approximately the same proportion of total slaves in both the New and Old South (4.5 percent).

TABLE 25

CROP OUTPUT, WEALTH, ACREAGE AND POPULATION BY SLAVEHOLDING CLASSES
AS A PERCENT OF REGIONAL TOTALS
OLD SOUTH VS NEW SOUTH-1860

Item	Region	Size of Slaveholding							
		0	1-4	5-9	10-19	20-49	50-99	100-199	200 & Larger[a]
No. of Farms	OS	41.3	18.2	14.0	13.1	10.0	2.4	.8	.1
	NS	55.5	14.4	9.2	9.7	8.6	.2	.1	.0
Val. of Pers. Prop.	OS	3.1	6.0	10.8	19.8	32.6	18.0	8.8	.9
	NS	5.5	6.3	9.2	19.5	34.2	17.7	7.5	.1
Imp. Acreage	OS	13.1	9.8	12.5	17.3	27.0	12.7	7.1	.5
	NS	19.6	9.0	8.9	16.6	26.0	12.3	7.0	.6
Unimp. Acreage	OS	16.9	11.7	13.3	16.7	27.4	10.1	3.4	.0
	NS	25.5	14.4	10.6	15.3	18.3	11.0	4.1	.7
Val. of Farm	OS	12.0	9.6	12.7	18.0	27.4	12.9	6.9	.5
	NS	11.1	7.4	6.7	14.2	25.4	19.6	14.2	1.2
Horses	OS	26.0	15.9	15.2	16.1	17.0	6.9	2.5	.3
	NS	35.8	12.8	10.4	15.5	18.8	4.7	1.6	.3
Asses & Mules	OS	8.5	7.8	10.8	18.8	30.3	15.2	8.2	.4
	NS	16.4	6.1	7.3	16.5	27.8	15.0	9.6	1.4
Swine	OS	21.5	13.7	13.6	16.5	22.0	8.8	3.5	.4
	NS	32.0	12.5	10.7	15.1	19.9	6.4	3.0	.4
Val. of Livestock	OS	16.3	11.5	13.5	18.1	23.7	11.8	4.9	.3
	NS	23.7	10.7	9.7	18.5	22.0	9.7	5.1	.5
Corn	OS	16.3	11.4	13.4	17.6	25.5	9.9	5.5	.4
	NS	22.5	9.9	9.1	15.8	25.2	11.2	5.4	.9

TABLE 25 (Continued)

Item	Region	0	1-4	5-9	10-19	20-49	50-99	100-199	200 & Larger[a]
		Size of Slaveholding							
Tobacco	OS	13.3	14.8	30.7	11.3	29.3	0	.6	0
	NS	39.7	22.2	19.9	2.0	15.9	.3	0	0
Cotton	OS	8.4	7.8	8.8	16.5	28.6	17.7	12.0	.3
	NS	8.9	5.2	6.4	14.5	32.1	18.3	13.4	1.0
Peas & Beans	OS	19.3	10.8	11.9	14.6	22.2	10.8	10.2	.2
	NS	22.8	10.5	7.6	17.6	27.1	10.6	3.1	.6
Irish Potatoes	OS	19.5	10.9	11.8	26.3	17.8	10.9	2.4	.3
	NS	30.0	10.3	9.4	16.7	20.3	8.6	3.0	1.7
Sweet Potatoes	OS	22.8	11.5	14.2	15.6	20.1	9.3	6.4	.1
	NS	26.3	11.3	8.1	15.4	22.8	11.6	3.3	1.1
Cane Sugar	OS	1.7	.8	.8	2.5	92.4	.8	.8	0
	NS	0	0	0	12.4	39.4	6.4	0	41.8
Molasses	OS	20.9	12.8	5.3	10.6	42.3	4.8	3.4	0
	NS	2.4	.7	.2	15.9	36.8	10.5	0	33.5
Val. of Home Manu.	OS	31.7	16.1	12.9	15.2	14.2	6.0	3.7	.1
	NS	50.1	12.1	8.3	12.0	13.5	2.6	1.4	0
Val. of An. Slau.	OS	19.7	11.8	13.3	23.6	21.1	7.0	3.1	.4
	NS	29.2	11.5	9.3	15.5	22.0	8.9	3.2	.4
Free Males	OS	43.2	18.1	13.6	12.2	9.3	2.7	.7	.1
	NS	57.9	13.9	9.1	9.0	7.7	1.8	.6	.0
Free Females	OS	43.7	17.2	13.1	12.9	9.2	2.7	1.1	.1
	NS	58.3	14.5	8.3	9.1	7.5	1.7	.5	.0

TABLE 25 (Continued)

Item	Region	Size of Slaveholding: 0	1-4	5-9	10-19	20-49	50-99	100-199	200 & Larger[a]
Slave Males	OS	0	4.4	10.2	20.4	34.8	18.4	10.6	1.2
	NS	0	3.8	8.5	19.4	36.5	17.9	12.5	1.4
Slave Females	OS	0	4.7	11.0	20.2	34.3	18.1	10.4	1.3
	NS	0	5.1	9.0	18.6	36.2	18.0	11.7	1.4
Total Population[b]	OS	17.1	9.7	11.7	17.3	24.6	12.2	6.7	.8
	NS	26.5	8.9	8.8	14.4	23.2	10.6	6.8	.8

[a]The sample included no slaveholdings larger than 499 slaves and only one slaveholding in each region in the 200-499 category.

[b]Total of four preceeding items.

Source: Calculated from 1860 manuscript census sample.

Clearly, the farmers operating smaller farms owned a greater proportion of wealth and produced a greater percentage of crops in the New South than did their counterparts in the Old South. Also, nonslaveholders were relatively more significant owners of wealth and producers of crops in the New South than in the Old South. But, when we consider only farmers who actually owned slaves, these differences disappear. Thus, we can conclude that the small landowner and nonslaveholder accounted for a greater share of wealthholdings and crop production in the New South than in the Old in 1860 but we cannot reach this same conclusion regarding the smaller slaveholders.

<u>New South vs Alluvial vs Old South</u>

The second comparison of the role of the small farmers in the New and Old South utilizes a subsample of the 5,229 farm 1860 sample. This subsample consists of the 964 sample farms which are selected from Gray's seventy-four major cotton producing counties. Of these, 452 farms are selected from counties which Gray designates as Old South; 208 from counties designated Alluvial; and 304 from counties designated as belonging to the New South.

Before examining the role of the small farmer in each of these regions, we should first determine whether the subsamples are representative of the regions from which they have been selected. Once again, we begin by comparing the mean crop outputs on the sample farms with the crop outputs per farm in the appropriate universe. The results of this test are summarized in Table 26. Using a five percent level of significance, only one of the Old South sample means (sweet potatoes) is significantly different from the universe mean. In the Alluvial region,

TABLE 26

TESTS OF THE 1860 SAMPLE BY REGION
OLD SOUTH VS ALLUVIAL VS NEW SOUTH
(Acreage, Value of Farm, Crop Outputs, Etc.)

	Old South Regions	Alluvial Regions	New South Regions
Number of farms in sample:	452	208	304
Values of t-statistics:			
Improved acres	1.98	-.02	-3.08
Unimproved acres	.99	.10	-1.83
Value of farm	.38	.98	-3.52
Value of farm machinery	1.34	1.31	-2.19
Swine	.75	-1.39	-.37
Value of livestock	.82	-.10	-1.30
Bushels of corn	1.08	.02	-4.43
Bales of cotton	.58	-2.25	-2.60
Bushels of peas & beans	-.05	-.67	.49
Bushels of Irish potatoes	-.11	.54	.01
Bushels of Sweet potatoes	2.15	-.63	.13
100 pounds of cane sugar	-.30	1.31	--
Gallons of molasses	-.33	1.34	.32
Value of home manufactures	-.45	-3.93	-.78
Value of animals slaughtered	.80	-.37	-3.97

Source: Calculated from 1860 manuscript census sample and U.S. Bureau of the Census, *Eighth Census of the United States: 1860, Agriculture*.

two of the sample means are significantly different from the universe values. However, in the New South region, six of the sample means are significantly different from the universe values. More significant for our purposes, perhaps, is the fact that ten of the sample means are less than the universe values. Further, none of the positive t-statistics exceeds .50. Thus, it appears that the New South sample is biased in the direction of mean crop outputs per farm which are significantly lower than the universe values.

The second test of the samples, a comparison of the size distribution of farms in the samples with those in the respective universes, confirms the suspicion that the New South sample may be biased in the direction of smaller farms. (See Table 27.) Using a five percent rejection level, the size distribution of farms in the New South sample is found to be significantly different from that of the New South universe. Also, the size distribution of farms in the Old South sample is significantly different from that of the Old South universe. However, for the Old South, the sample is biased in the direction of larger farms. This bias is most obvious in the 1000 acres and larger category. Luck has not completely deserted us, however, for the size distribution of farms in the Alluvial sample is virtually identical to that of the universe.

TABLE 27

TESTS OF THE 1860 SAMPLE BY REGION
OLD SOUTH VS ALLUVIAL VS NEW SOUTH
(Size Distribution of Farms)

Size of Farm (Imp. Acreage)	Old South Regions Observed	Old South Regions Expected	Alluvial Regions Observed	Alluvial Regions Expected	New South Regions Observed	New South Regions Expected
3-9 acres	3	3.4	4	4.4	9	7.7
10-19 acres	10	10.9	16	17.6	32	22.0
20-49 acres	67	69.0	34	39.5	76	84.6
50-99 acres	85	96.4	39	31.6	70	67.0
100-499 acres	210	213.2	79	78.0	101	94.3
500-999 acres	40	37.2	26	26.6	7	16.6
1000 acres & up	29	13.9	9	9.3	4	6.8
Total*	444		207		299	
	x^2(df=6) = 18.20 p = .005		x^2(df=6) = 2.72 p = .84		x^2(df=6) = 12.95 p = .05	

*Farms of less than three acres are not included in the Table. The number of sample farms having less than three acres of improved land is:

Old South - 8
Alluvial - 1
New South - 5

Source: Calculated from 1860 manuscript census sample and U.S. Bureau of the Census, Eighth Census of the United States: 1860, Agriculture.

The third test of the regional samples is a comparison of the size distribution of slaveholdings in each of the samples with those of their respective universes. (See Table 28.) This test yields a rather unexpected result. It has been observed earlier that, due to the fact that the sample is selected entirely from rural regions, we expect the samples to be deficient in the smaller slaveholding categories. This is indeed the case for the samples selected from the Old South and the Alluvial

TABLE 28

TESTS OF THE 1860 SAMPLE BY REGION
OLD SOUTH VS ALLUVIUM VS NEW SOUTH
(Size Distribution of Slaveholdings)

Size of Slaveholding		Old South Regions		Alluvial Regions		New South Regions	
		Observed	Expected	Observed	Expected	Observed	Expected
1	slave	23	40.7	6	20.0	13	22.9
2	slaves	12	28.4	9	13.6	14	15.6
3	slaves	10	23.6	6	10.2	6	11.7
4	slaves	15	20.1	4	8.7	7	10.6
5	slaves	25	17.8	7	7.8	8	8.5
6	slaves	10	16.8	6	6.4	13	7.7
7	slaves	11	14.0	2	5.4	7	6.3
8	slaves	21	13.3	8	5.1	6	6.0
9	slaves	9	11.3	4	3.9	4	4.9
10-14	slaves	50	41.0	13	15.5	23	18.3
15-19	slaves	29	24.7	15	9.9	16	10.8
20-29	slaves	32	27.7	20	13.7	16	11.3
30-39	slaves	22	15.2	11	8.0	9	6.7
40-49	slaves	14	8.3	14	5.5	5	4.2
50-69	slaves	17	8.8	9	7.5	4	3.9
70-99	slaves	14	5.3	7	5.7	2	3.1
100-199	slaves	6	2.6	10	4.0	2	2.0
200-299	slaves	--	.2	--	.6	--	.3
300-499	slaves	--	.1	1	.1	--	.1
500-999	slaves	--	--	--	--	--	--
1000 & up		--	--	--	--	--	--
		320		152		155	
		x^2(df=16) = 73.11 p = .0005		x^2(df=16) = 48.77 p = .0005		x^2(df=15) = 16.88 p = .33	
Farms without Slaves		132		56		149	

Source: Calculated from 1860 manuscript census sample and U.S. Bureau of the Census, Eighth Census of the United States: 1860, Agriculture.

regions. However, the New South sample, which at first glance is deficient in the smaller slaveholding classes, actually has a size distribution of slaveholdings which is not significantly different from that of the universe (which consists of both urban and rural slaveholdings). Thus, it again appears that the New South sample is biased in the direction of smaller farms and slaveholdings.

On the basis of the above tests, it is questionable whether the samples here being considered are representative of their respective universes. Were the world one of unlimited resources, it would probably be advisable to select new samples from the Old South and New South regions. (Of course, the new samples might also prove to be unrepresentative.) However, given our limited resources, we shall go ahead and use these regional samples bearing in mind that the Old South sample has an abundance of very large farms while the New South sample appears to consist of a collection of farms which are smaller than we had hoped. (Unfortunately, the biases in these samples are such as to prejudice our test in the direction of rejecting the first part of our hypothesis. Because the earlier data suggests that the hypothesis should be rejected, these biases are precisely the ones which are the most damaging to our case.)

The size distributions of sample farms in the Old South, Alluvial and New South regions are compared in Table 29. Smaller farms were relatively more common in the newer regions (New South and Alluvial) than in the Old South region. Only 19.5 percent of Old South farms contained less than fifty improved acres while 26.4 percent of the Alluvial farms and 40.1 percent of the New South farms were this size. If farms of less than one hundred improved acres are contrasted, the same

TABLE 29

SIZE DISTRIBUTION OF FARMS (IMPROVED ACREAGE)
OLD SOUTH, ALLUVIAL, NEW SOUTH - 1860

Farm Sizes	Old South		Alluvial		New South	
Less than 3 acres	8	(1.8%)	1	(.5%)	5	(1.6%)
3-9 acres	3	(.7%)	4	(1.9%)	9	(3.0%)
10-19 acres	10	(2.2%)	16	(7.7%)	32	(10.5%)
20-49 acres	67	(14.8%)	34	(16.3%)	76	(25.0%)
50-99 acres	85	(18.8%)	39	(18.7%)	70	(23.0%)
100-199 acres	113	(24.9%)	37	(17.8%)	57	(18.7%)
200-299 acres	51	(11.2%)	15	(7.2%)	21	(7.0%)
300-499 acres	46	(10.2%)	27	(13.0%)	23	(7.5%)
500-999 acres	40	(8.8%)	26	(12.5%)	7	(2.3%)
1000 acres & larger	29	(6.4%)	9	(4.3%)	4	(1.3%)
Total	452		208		304	

Old South: $x^2(df = 9) = 25.2$, $p < .005$

New South: $x^2(df = 9) = 36.2$, $p < .005$

Alluvial: $x^2(df = 9) > 23.6$, $p < .005$

Source: Calculated from 1860 manuscript census sample.

result obtains. Only 38 percent of the Old South farms contained less than one hundred acres. In the Alluvial lands this percentage was over 45 percent. In the New South an overwhelming majority of farms (over 63 percent) contained less than one hundred improved acres.

At this point, the reader may begin to recall our observation that the New South sample contains an excessive number of small farms while the Old South sample includes too many large farms. These biases indeed appear to cast some doubt on the findings presented in Table 29. A closer look at Table 27, however, indicates that the biases in the samples are not sufficient to explain the very great differences in the distributions of farm sizes between the three regions. This point may be conclusively demonstrated by calculating the percentage of farms of less than fifty acres in each of the universes and the percentage which are less than one hundred acres using data from the published census. (Universe data is available only for farms of three acres and larger. Thus, farms of less than three acres cannot be considered in the following discussion.) Using data from the "expected" columns in Table 27, (calculated from the published census) we find that approximately 19 percent of the farms in the Old South universe, 30 percent of the farms in the Alluvial universe, and 38 percent of the farms in the New South universe consisted of less than fifty improved acres. If universe farms of less than one hundred acres are considered the percentages are: Old South--40 percent, Alluvial--45 percent, and New South 61 percent. Thus, the distributions based on our sample are not sufficiently different from the universe distributions to cause the very great interregional differences.

Having demonstrated that Table 29 presents a reliable picture of the regional differences, we conclude that small farms were relatively more common in the New South than in the Old South in 1860. Further, smaller farms were a greater proportion of Alluvial farms than of Old South farms but the differences between Alluvial and Old South were not so great as between New and Old South.

As may be seen in Table 30, nonslaveholders were much more common in the New South than in either the Alluvial or Old South regions. In fact, approximately one-half of the farmers in the New South region did not own a single slave. In the other regions only one-fourth of the farmers were slaveless. The size distributions of slaveholdings in the Old and New South are significantly different at the .01 level as are the Alluvial and New South distributions. However, the Old South and Alluvial distributions are not significantly different even at the .05 level.

If nonslaveholders are omitted from the size distributions of slaveholdings, the findings are considerably altered. (See Table 31.) First, the New South and Old South distributions, with nonslaveholders omitted, are not significantly different. However, the New South led the Old South in the percentage of slaveholdings of 1-4, 5-9, and 10-19 slaves. The respective percentages were: 1-4 slaves: Old South - 18.8, New South - 25.8; 5-9 slaves: Old South - 23.7, New South - 24.6; and 10-19 slaves: Old South - 24.7, New South - 25.1. Thus, over seventy-five percent of New South slaveholdings were of less than twenty slaves. Only sixty-seven percent of the Old South slaveholdings contained less than twenty slaves. The size distribution of Alluvial slaveholdings is also not significantly different from the Old South distribution at the .05 level. However, in the Alluvial regions, slaveholdings clearly appear to have been larger than in the Old South. In the Alluvial, 7.3 percent of all slaveholdings were larger than one hundred slaves as compared to

TABLE 30

DISTRIBUTION OF SLAVEHOLDINGS
OLD SOUTH, ALLUVIAL, NEW SOUTH - 1860

Slaveholdings		Old South		Alluvial		New South	
0	slaves	132	(29.2%)	56	(26.9%)	149	(49.0%)
1	slave	23	(5.1%)	6	(2.9%)	13	(4.3%)
2	slaves	12	(2.6%)	9	(4.3%)	14	(4.6%)
3	slaves	10	(2.2%)	6	(2.9%)	6	(2.0%)
4	slaves	15	(3.3%)	4	(2.0%)	7	(2.3%)
5	slaves	25	(5.5%)	7	(3.4%)	8	(2.6%)
6	slaves	10	(2.2%)	6	(2.9%)	13	(4.3%)
7	slaves	11	(2.4%)	2	(.9%)	7	(2.3%)
8	slaves	21	(4.6%)	8	(3.8%)	6	(2.0%)
9	slaves	9	(2.0%)	4	(2.0%)	4	(1.3%)
10-14	slaves	50	(11.0%)	13	(6.2%)	23	(7.5%)
15-19	slaves	29	(6.4%)	15	(7.2%)	16	(5.3%)
20-29	slaves	32	(7.0%)	20	(9.6%)	16	(5.3%)
30-39	slaves	22	(4.8%)	11	(5.3%)	9	(3.0%)
40-49	slaves	14	(3.1%)	14	(6.7%)	5	(1.6%)
50-69	slaves	17	(3.7%)	9	(4.3%)	4	(1.3%)
70-99	slaves	14	(3.1%)	7	(3.3%)	2	(.7%)
100-199	slaves	6	(1.3%)	10	(4.8%)	2	(.7%)
200-299	slaves	---	---	---	---	---	---
300-499	slaves	---	---	1	(.5%)	---	---
500-999	slaves	---	---	---	---	---	---
1000 slaves & up		---	---	---	---	---	---
	Total	452		208		304	

$x^2(df = 17) = 25.86$, $p = .08$ $\qquad$ $x^2(df = 17) > 35.7$, $p < .005$

$x^2(df = 17) > 35.7$, $p < .005$

Source: Calculated from 1860 manuscript census sample.

TABLE 31

DISTRIBUTION OF SLAVEHOLDINGS (EXCLUDING NONSLAVEHOLDERS)
OLD SOUTH, ALLUVIAL, NEW SOUTH - 1860

Slaveholdings		Old South		Alluvial		New South	
1	slave	23	(7.2%)	6	(3.9%)	13	(8.4%)
2	slaves	12	(3.8%)	9	(5.9%)	14	(9.0%)
3	slaves	10	(3.1%)	6	(3.9%)	6	(3.9%)
4	slaves	15	(4.7%)	4	(2.6%)	7	(4.5%)
5	slaves	25	(7.8%)	7	(4.6%)	8	(5.2%)
6	slaves	10	(3.1%)	6	(3.9%)	13	(8.4%)
7	slaves	11	(3.4%)	2	(1.3%)	7	(4.5%)
8	slaves	21	(6.6%)	8	(5.3%)	6	(3.9%)
9	slaves	9	(2.8%)	4	(2.6%)	4	(2.6%)
10-14	slaves	50	(15.6%)	13	(8.6%)	23	(14.8%)
15-19	slaves	29	(9.1%)	15	(9.9%)	16	(10.3%)
20-29	slaves	32	(10.0%)	20	(13.2%)	16	(10.3%)
30-39	slaves	22	(6.9%)	11	(7.2%)	9	(5.8%)
40-49	slaves	14	(4.4%)	14	(9.2%)	5	(3.2%)
50-69	slaves	17	(5.3%)	9	(5.9%)	4	(2.6%)
70-99	slaves	14	(4.4%)	7	(4.6%)	2	(1.3%)
100-199	slaves	6	(1.9%)	10	(6.6%)	2	(1.3%)
200-299	slaves	---	---	---	---	---	---
300-499	slaves	---	---	1	(.7%)	---	---
500-999	slaves	---	---	---	---	---	---
1000 slaves & up		---	---	---	---	---	---
Total		320		152		155	

$x^2(df = 16) = 25.00$
$p = .07$

$x^2(df = 16) = 28.92$
$p = .025$

$x^2(df = 16) = 20.07$
$p = .20$

Source: Calculated from 1860 manuscript census sample.

only 1.9 percent in the Old South. This difference is all the more striking if we recall that the Old South sample appears to be biased in the direction of larger farms. The differences between the Alluvial and New South distributions are more striking and are statistically significant at the .05 level. In the Alluvial, 7.3 percent of the slaveholdings contained one hundred slaves or more and almost 18 percent contained fifty slaves or more. In the New South region, these figures were 1.3 percent and 5.2 percent respectively. Over one-fourth of the New South slaveholdings contained less than five slaves. Approximately one-sixth of the Alluvial holdings contained less than five slaves.

Thus, we find that small land-owners were relatively more numerous in the newer regions--Alluvial and New South--than in the older regions. Also, small land-owners were a greater percentage of farmers in the New South than in the Alluvial. Further, nonslaveholdings were much more common in the New South than in either of the other regions. Finally, small slaveholdings may have been more common in the New South than in the other regions although this result is not statistically significant. Alluvial slaveholdings, on the other hand, tended to be larger than those of the other regions.

Again, the hypothesis that large planters led the migration into the Southwest appears questionable. However, we can begin to understand where this hypothesis may have originated. If historians have been rather myopic and focused on the Alluvial regions, the 1860 data may explain their conclusions.

Next, we take a closer look at each of the samples and attempt to determine what contribution the small farms made to their respective regions. Crop outputs, wealth, acreage and population by acreage class expressed as a percent of regional totals are given in Table 32. In every category, the share of the total New South property owned or crops produced by farmers with less than twenty acres exceeded the share of total Old South or Alluvial property owned or crops produced by Old South or Alluvial farmers operating farms of less than twenty acres of improved land. This same conclusion obtains if we compare the contributions of farmers with less than fifty acres in each of the three regions (with the exception of tobacco production). The differences between the New South and the other regions are particularly striking in the improved acreage category and in the production of corn, cotton, and peas and beans. Also, the share of livestock owned by small farmers in the New South is particularly striking. In the New South region, over thirty-five percent of the rural free population lived on farms with less than fifty improved acres. In the Old South, this figure was less than twenty percent and in the Alluvial region it was approximately twenty-five percent. Finally, a greater percentage of rural slaves lived on farms of less than fifty acres in the New South than in either the Alluvial or Old South regions.

The percentages of total Alluvial property owned and crop output produced by farmers with less than twenty acres were, as a general rule, only slightly higher than the percentages of property owned and crops produced on farms of less than twenty acres in the Old South. Exceptions were swine and corn production where the share produced by small farmers in the Alluvial region exceeded that in the Old South region. Further, over ten percent of the rural free population in the Alluvial region lived on farms with less than twenty acres as compared to less than four percent of the rural free population in the Old South. If farms of less than fifty improved acres are contrasted, the role of the small farmers

TABLE 32

CROP OUTPUT, WEALTH, ACREAGE AND POPULATION BY ACREAGE CLASSES
AS A PERCENT OF REGIONAL TOTALS
OLD SOUTH VS ALLUVIAL VS NEW SOUTH - 1860

		Farm Sizes (Improved Acreage)							
Item	Region	Less Than 20	20-49	50-99	100-199	200-299	300-499	500-999	1000 & Larger
No. of Farms	OS	4.7	14.8	18.8	25.0	11.3	10.2	8.8	6.4
	All	10.1	16.3	18.8	17.8	7.2	13.0	12.5	4.3
	NS	15.1	25.0	23.0	18.8	6.9	7.6	2.3	1.3
Val. of Pers. Prop.	OS	.2	2.4	3.3	13.5	11.1	20.5	22.3	26.7
	All	.2	1.3	4.5	12.5	8.0	24.6	33.7	15.1
	NS	2.8	5.8	8.7	16.7	16.1	23.8	12.4	13.7
Imp. Acreage	OS	.1	1.8	5.0	13.3	10.4	14.4	22.0	33.1
	All	.5	1.9	5.0	9.4	6.4	18.5	32.8	25.4
	NS	1.3	6.4	12.2	18.7	12.8	22.3	12.4	13.8
Unimp. Acreage	OS	.4	4.3	7.6	15.7	9.7	16.0	24.2	22.1
	All	1.3	5.1	12.2	14.1	8.9	13.7	23.6	20.9
	NS	4.7	9.7	22.2	21.6	14.8	17.0	4.3	5.7
Val. of Farm	OS	.3	3.0	6.7	16.8	11.2	15.1	21.5	25.5
	All	.2	.8	3.1	8.9	4.1	15.7	33.7	33.4
	NS	2.1	5.6	12.3	16.2	21.9	18.5	12.6	10.8
Horses	OS	1.0	7.6	11.8	23.4	11.3	13.4	16.0	15.5
	All	3.9	6.6	10.8	15.1	5.4	17.8	24.6	15.7
	NS	9.0	15.2	28.1	19.2	8.0	9.9	5.8	4.8
Asses and Mules	OS	.2	1.8	3.6	13.1	10.0	17.3	22.5	31.5
	All	.6	1.5	2.1	9.5	5.0	19.1	30.7	31.5
	NS	1.8	5.6	9.5	17.0	17.3	25.2	9.7	13.9

TABLE 32 (Continued)

Item	Region	Farm Sizes (Improved Acreage) Less Than 20	20-49	50-99	100-199	200-299	300-499	500-999	1000 & Larger
Swine	OS	1.1	5.8	9.4	17.6	10.0	14.3	19.7	22.3
	All	4.3	6.9	10.1	15.0	7.7	22.0	21.4	12.5
	NS	4.9	16.2	16.1	22.7	10.6	19.5	7.6	2.3
Val. of Livestock	OS	.4	4.4	7.5	17.2	10.6	14.3	20.4	24.9
	All	1.6	3.4	6.3	11.5	12.0	19.1	27.5	18.6
	NS	5.3	9.3	17.5	17.5	14.4	18.8	9.0	8.3
Corn	OS	.4	3.5	7.3	15.4	9.4	16.0	21.3	26.5
	All	1.3	2.6	6.3	11.9	7.5	19.3	31.9	19.1
	NS	2.8	8.4	14.5	20.0	13.5	19.7	14.9	6.2
Tobacco	OS	0	71.4	0	0	28.6	0	0	0
	All	0	0	0	0	0	0	0	0
	NS	0	40.0	40.0	20.0	0	0	0	0
Cotton	OS	.1	1.7	4.8	13.5	7.7	13.6	22.6	35.9
	All	.2	.8	3.6	10.4	4.5	19.3	33.5	27.6
	NS	.8	5.1	7.3	15.9	20.4	23.0	15.8	11.7
Peas and Beans	OS	.9	6.2	8.2	16.3	10.0	17.2	20.5	20.6
	All	1.9	5.2	7.0	17.6	3.9	18.1	34.8	11.6
	NS	3.1	15.4	20.6	21.5	11.9	22.0	2.0	3.5
Irish Potatoes	OS	.3	4.9	9.8	20.4	18.2	20.3	8.5	17.6
	All	2.8	6.2	5.2	14.7	5.6	21.0	26.5	18.0
	NS	4.8	10.3	14.9	31.1	12.0	13.0	7.4	6.5
Sweet Potatoes	OS	1.0	5.2	7.8	17.0	9.8	11.2	19.5	28.5
	All	2.3	5.7	9.3	18.0	7.2	23.0	23.0	11.5
	NS	4.4	11.3	14.4	21.7	14.6	21.1	9.9	2.5

TABLE 32 (Continued)

Item	Region	Farm Sizes (Improved Acreage) Less Than 20	20-49	50-99	100-199	200-299	300-499	500-999	1000 & Larger
Cane Sugar	OS	0	0	0	20.0	0	0	40.0	40.0
	All	0	0	0	53.2	0	0	0	46.8
	NS	0	0	0	0	0	0	0	0
Molasses	OS	0	.9	.3	3.1	6.9	21.8	13.8	53.2
	All	0	0	0	56.1	0	.1	.1	43.7
	NS	0	11.7	0	88.3	0	0	0	0
Val. of Home Manu.	OS	3.0	10.6	10.1	21.2	9.9	13.5	13.9	17.5
	All	2.9	8.5	13.2	7.9	20.1	7.8	31.3	8.3
	NS	10.7	17.3	24.0	26.2	4.0	13.7	2.7	1.3
Val. of An. Slau.	OS	1.2	4.6	11.9	18.8	10.7	14.9	16.8	21.2
	All	2.4	6.7	10.2	13.0	7.8	23.7	24.9	11.4
	NS	6.8	12.8	13.6	22.1	10.2	22.5	9.4	2.6
Free Males	OS	3.5	13.9	19.2	25.8	11.3	11.8	8.6	6.0
	All	10.4	17.3	20.9	16.6	6.4	10.9	14.1	3.4
	NS	10.7	23.5	26.2	21.7	5.1	8.8	3.1	.9
Free Females	OS	3.7	15.2	18.8	24.1	11.6	10.2	8.6	7.8
	All	11.5	16.6	18.7	17.3	6.3	11.8	15.6	2.3
	NS	13.5	24.7	25.1	20.5	5.2	7.5	2.7	.7
Slave Males	OS	0	1.7	2.7	11.8	10.4	16.9	24.7	31.7
	All	.1	.5	3.7	9.9	6.3	16.8	32.2	30.4
	NS	.5	3.2	7.2	16.0	20.5	26.1	14.1	12.4

TABLE 32 (Continued)

Item	Region	Farm Sizes (Improved Acreage) Less Than 20	20-49	50-99	100-199	200-299	300-499	500-999	1000 & Larger
Slave Females	OS	.2	1.9	3.3	12.6	12.0	16.9	26.5	26.7
	All	.1	.9	3.6	9.6	5.8	18.2	32.4	29.2
	NS	1.4	3.6	8.0	16.8	20.2	23.3	15.2	11.5
Total Population*	OS	1.2	5.4	7.5	15.8	11.3	15.2	20.8	22.9
	All	2.0	3.6	6.5	11.1	6.1	16.4	29.2	25.1
	NS	5.6	12.1	15.2	18.4	14.0	17.8	9.7	7.3

*Total of four preceeding items.

Source: Calculated from 1860 manuscript census sample.

in the Alluvial region was even more similar to the role of their Old South counterparts. Again, however, the percentage of free population living on farms of less than fifty acres in the Alluvial region greatly exceeded that in the Old South.

These results should be qualified by recalling that the Old South sample is biased in the direction of larger farms and the New South sample in the direction of smaller farms. However, it should be pointed out these biases are confined to the under twenty acre category. (In the Old South universe 18.8 percent of the farms contained less than fifty acres. For the sample the figure was 18.0 percent. In the New South the figures were: universe--38.3, sample--39.2.) Thus, the conclusions based on farms of less than fifty acres are not only the results of biases in the samples. Additionally, it should be pointed out that there are approximately thirty percent too many New South farms of less than twenty acres and ten percent too few Old South farms in that category. A closer examination of Table 32 reveals that, if we are to make the appropriate adjustments to compensate for the biases in the data, our conclusions would not be substantially altered.

Crop outputs, wealth, acreage and population data by slaveholding class are expressed as a percentage of regional totals in Table 33. Nonslaveholders in the New South accounted for a greater share of the wealth owned and crops produced than did nonslaveholders in each of the other two regions. The same conclusion obtains if slaveholders owning one to four slaves are considered. The percentages of wealth owned and property produced by nonslaveholders and small slaveholders in the Alluvial region were generally lower than these percentages in either of the other regions. Conversely, the share of wealth owned and crops

TABLE 33

CROP OUTPUT, WEALTH, ACREAGE AND POPULATION BY SLAVEHOLDING CLASSES
AS A PERCENT OF REGIONAL TOTALS
OLD SOUTH VS ALLUVIAL VS NEW SOUTH - 1860

Item	Region	Size of Slaveholding							
		0	1-4	5-9	10-19	20-49	50-99	100-199	200 & Larger[a]
No. of Farms	OS	29.2	13.3	16.8	17.5	15.0	6.9	1.3	0
	All	26.9	12.0	13.0	13.5	21.6	7.7	4.8	.5
	NS	49.0	13.2	12.5	12.8	9.9	2.0	.7	0
Val. of Pers. Prop.	OS	1.4	3.1	7.2	15.6	35.0	29.8	7.9	0
	All	1.2	1.7	6.3	12.1	35.7	27.2	15.2	.7
	NS	4.3	6.5	10.7	22.8	34.8	12.3	8.6	0
Imp. Acreage	OS	6.7	4.9	11.0	16.6	29.1	23.6	8.1	0
	All	3.5	3.1	5.3	9.7	30.8	20.6	22.2	4.6
	NS	15.4	7.7	10.2	19.3	29.2	11.7	6.5	0
Unimp. Acreage	OS	10.1	5.5	10.3	13.5	29.6	26.4	4.6	0
	All	7.4	4.6	12.2	12.6	23.9	14.2	18.6	7.5
	NS	23.1	11.1	11.6	25.4	22.7	4.0	2.1	0
Val. of Farm	OS	6.9	5.3	11.3	17.1	29.2	24.7	5.5	0
	All	2.1	1.9	3.2	6.8	21.4	19.9	39.4	5.2
	NS	10.5	6.3	7.8	23.0	29.0	12.1	11.3	0
Horses	OS	15.7	8.8	17.1	16.6	21.3	16.8	3.6	0
	All	12.0	6.5	9.2	13.5	25.3	16.5	12.3	4.6
	NS	33.1	12.0	17.6	16.0	13.9	5.5	2.0	0
Asses & Mules	OS	3.7	3.6	8.8	18.1	30.4	26.3	9.1	0
	All	2.0	1.6	3.1	7.0	29.8	20.1	28.5	7.8
	NS	8.6	4.1	9.0	23.4	33.1	15.4	6.4	0

TABLE 33 (Continued)

Item	Region	0	1-4	5-9	10-19	20-49	50-99	100-199	200 & Larger[a]
		Size of Slaveholding							
Swine	OS	11.9	7.3	11.5	18.6	23.3	22.3	5.1	0
	All	13.6	7.7	8.0	13.8	26.8	15.2	10.9	4.0
	NS	22.1	9.5	10.7	24.4	23.6	8.0	1.7	0
Val. of Livestock	OS	8.8	6.3	12.5	17.0	24.9	24.2	6.3	0
	All	6.1	4.1	5.1	16.7	27.3	17.4	19.0	4.3
	NS	18.9	8.0	13.0	22.5	23.5	8.8	5.3	0
Corn	OS	7.7	4.9	11.2	15.4	31.2	21.4	8.4	0
	All	5.4	4.2	5.7	9.3	34.4	18.0	15.7	7.3
	NS	16.1	7.6	9.4	23.5	25.6	10.0	7.7	0
Tobacco	OS	71.4	0	28.6	0	0	0	0	0
	All	0	0	0	0	0	0	0	0
	NS	96.0	4.0	0	0	0	0	0	0
Cotton	OS	4.6	4.2	8.7	15.2	27.1	28.0	12.0	0
	All	2.1	1.9	3.7	7.6	26.5	22.5	30.6	5.2
	NS	7.8	4.3	5.4	22.1	31.2	19.8	9.4	0
Peas and Beans	OS	10.8	7.2	9.4	11.0	30.7	23.8	7.2	0
	All	7.8	5.1	8.5	7.7	38.4	21.4	7.0	4.0
	NS	22.2	11.8	13.7	19.4	29.6	.2	3.1	0
Irish Potatoes	OS	10.1	8.3	11.0	28.6	24.6	15.9	1.5	0
	All	12.2	3.9	6.5	7.2	25.5	22.1	12.2	10.5
	NS	20.3	8.6	11.5	34.2	16.3	5.1	4.1	0
Sweet Potatoes	OS	9.7	5.8	11.8	14.2	24.6	24.2	9.7	0
	All	12.7	3.4	6.6	7.5	35.6	18.3	5.5	10.4
	NS	21.1	9.5	7.8	21.8	27.5	10.6	1.6	0

TABLE 33 (Continued)

Item	Region	Size of Slaveholding 0	1-4	5-9	10-19	20-49	50-99	100-199	200 & Larger[a]
Cane Sugar	OS	0	0	0	20.0	0	40.0	40.0	0
	All	0	0	0	9.0	44.2	0	0	46.8
	NS	0	0	0	0	0	0	0	0
Molasses	OS	.9	1.6	0	15.6	22.3	29.8	29.8	0
	All	.0	.0	0	9.6	46.7	0	0	43.6
	NS	3.7	28.4	67.9	0	0	0	0	0
Val. of Home Manu.	OS	20.8	8.8	11.9	11.5	21.8	22.6	2.7	0
	All	15.9	9.2	7.5	7.7	36.0	6.3	17.4	0
	NS	41.6	14.7	10.8	10.8	22.2	0	0	0
Val. of An. Slau.	OS	12.7	7.0	13.3	17.8	26.3	18.3	4.7	0
	All	11.3	5.8	7.2	13.0	31.0	18.5	8.8	4.4
	NS	24.8	8.6	9.8	21.1	27.6	5.7	2.4	0
Free Males	OS	29.9	13.3	17.7	16.4	14.6	7.1	1.0	0
	All	34.1	8.7	13.0	11.7	18.8	9.8	3.8	.2
	NS	50.2	13.0	12.8	11.9	10.0	1.4	.8	0
Free Females	OS	30.7	11.9	14.6	18.4	13.6	8.0	2.8	0
	All	30.3	10.1	12.6	12.2	22.5	9.5	2.3	.6
	NS	50.5	14.1	12.9	12.1	8.0	1.3	1.1	0
Slave Males	OS	0	2.0	6.9	15.6	31.1	33.8	10.6	0
	All	0	1.0	4.1	8.9	28.4	22.6	28.3	6.8
	NS	0	2.8	9.2	22.4	38.4	16.6	10.6	0

TABLE 33 (Continued)

Item	Region	Size of Slaveholding							
		0	1-4	5-9	10-19	20-49	50-99	100-199	200 & Larger[a]
Slave Females	OS	0	2.1	8.2	16.6	31.8	30.7	10.7	0
	All	0	1.3	3.4	8.4	31.0	21.2	28.1	6.7
	NS	0	4.4	11.5	21.4	36.1	17.8	9.0	0
Total Population[b]	OS	8.5	5.0	10.0	16.5	26.5	25.2	8.2	0
	All	5.6	2.6	5.3	9.2	28.1	19.7	23.8	5.6
	NS	21.1	7.8	11.4	17.7	25.4	10.5	6.0	0

[a]The sample included no slaveholdings larger than 499 slaves and only one farm in the 200-499 category (Alluvium).

[b]Total of four preceeding items.

Source: Calculated from 1860 manuscript census sample.

produced by the larger slaveholders were greatest in the Alluvial region. It is interesting to note that over half of the rural free population in the New South lived on farms without slaves. Approximately thirty percent of the free population in each of the other regions lived on slaveless farms.

Thus, nonslaveholders made a relatively greater contribution to the rural economy of the New South region than they did in either of the other regions. In the Alluvial region, larger slaveholders appear to have been more common than in the Old or New South regions. In both the Old South and Alluvial regions over thirty percent of the rural free population lived on farms without slaves. Even more striking, however, the majority of the rural free population in the New South lived on farms without a single slave.

CONCLUSIONS

The above analysis of the 1860 sample certainly casts doubt on the hypothesis that large planters led the migration into the newer regions. When the 5229 farm sample is divided into two parts, Old South and New South, we observe three interesting results.

First, operators of small farms were a greater percentage of farmers in the New South than in the Old South. These smaller farmers held a greater percentage of wealth in the New South than did their counterparts in the Old South. Excepting sugar and molasses, they produced a greater share of the crops in the New South. Finally, the percentage of the rural free population living on farms of less than twenty acres was two and one-half times as great in the New South as in the Old South. Clearly, the small farmer was more common in the New South than in the Old South. Thus, unless one is willing to argue that the large planters who moved to the newer regions operated smaller farms than they had operated in the Old South, our hypothesis appears questionable. Actually, a closer examination of Table 24 indicates that it is highly unlikely that the small farmers in the New South were large planters from the Old South who merely reduced the size of their acreage holdings when they moved west. If this were the case, one would expect to find the proceeds of the land sale reflected in a higher value of personal property. In fact, the percentage of farmers with less than twenty acres _and_ the percentage of personal property owned by these same farmers were _each_ 2.7 times greater in the New South than in the Old South. Clearly, the wealth holdings of small farmers in the New South provide _no_ evidence that the small farmers in the Southwest had previously been large planters in the Southeast.

Second, nonslaveholdings were much more common in the New South. These nonslaveowners held a greater percentage of wealth and produced a larger share of the crops in the New South than in the Old South. Further, almost sixty percent of the rural free population in the New South lived on slaveless farms as compared with approximately forty-three percent in the Old South. Again, we find no evidence that the large planter, slaves in tow, led the migration.

Finally, if we just consider persons who actually owned slaves, we observe some tendency for smaller slaveholders to have been more common in the New South. However, their greater numbers and their contribution to the economy do not appear to have been significantly different in the New and Old South.

When we analyze the three 1860 subsamples--Old South, Alluvial, and New South--we find somewhat the same results. However, by examining the characteristics of the Alluvial region independently from the New South, we are able to shed some light on the source of our hypothesis.

First, we find that small farms were relatively more numerous in the New South than in either of the other regions and more common in the Alluvial than in the Old South. The smaller farmers in the New South owned a greater percentage of the regional wealth and produced a larger share of the crops than did their counterparts in the other two regions. Over thirty-five percent of the rural free population lived on farms with less than fifty acres in the New South. This was in contrast with less than twenty percent in the Old South and approximately twenty-five percent in the Alluvial. Again, we find no evidence that large planters led the migration into the Southwest. However, we do find evidence that many large farmers may have migrated to the Alluvial regions.

Second, we find that nonslaveholders were much more common in the New South than in the other two regions. Approximately one-half of the farmers in the New South region were slaveless as compared with less than thirty percent in the Old South and Alluvial. Nonslaveholders accounted for a greater percentage of wealth owned and crops produced in the New South than in the other regions; in the Alluvial, nonslaveholders accounted for a smaller share. If nonslaveholders are omitted from the analysis, we find that the New South and Old South distributions were not significantly different. But, large slaveholdings appear to have been more common in the Alluvial than in the other regions. Further, the share of wealth owned and crops produced by large slaveholders was greatest in the Alluvial region.

Thus, our comparison of these regions leads us to the conclusion that there is no evidence that large planters led the migration into the

New South. However, they may have led the way into the Alluvial regions or, perhaps, purchased land in the Alluvial from the smaller farmers. (More on this latter possibility in Chapter VI.) Evidently, the Alluvial region in 1860 had larger slaveholdings _but_ smaller farms than did the Old South. This may reflect the fact that, on the richer Alluvial farms, size of operation was limited by the number of slaves which could be purchased and organized rather than by the number of acres which could be worked. At least, the factors of production appear to have been employed as economic theory would suggest they should.

From the above observations based on the 1860 sample, we can formulate three tentative conclusions which are subject to modification after we consider the 1850 data. First, if the entire New South region is compared to the entire Old South, there is no evidence that large planters led the migration into the new lands in the Southwest. In fact, relatively few large planters had arrived in the region by 1860. Second, if we examine the data from the Old South, Alluvial, and New South regions, we again find no evidence that larger planters led the migration into the Southwest. However, we find that in 1860 larger slaveholders were more common in the Alluvial than in the Old or New South.

CHAPTER VI

TESTING THE HYPOTHESIS--1850

The results presented in Chapter IV suggest that by 1860 small farmers and nonslaveholders were relatively more common in the New South than in the Old South. Also the percentage of wealth owned and crops produced by these smaller farmers and nonslaveholders was greater in the New South. The possibility exists, however, that the small farmers were late-comers. That is, the small farmers may have followed large planters into the Southwest and followed in such great numbers as to become more predominant in the Southwest than in the Southeast by 1860.

In this chapter we will examine this possibility by comparing the size distribution of farms and slaveholdings in the Old South, Alluvial and New South regions in 1850. Further, we will compare the percentage of wealth owned and crops produced by the smaller farmers and slaveholders in each of the regions.

DEFINITION OF REGIONS

The 897-farm 1850 sample is selected only from the seventy-four counties described by Gray as the major cotton producing counties in 1859. The regional breakdown employed in this chapter is identical to the second regional breakdown used in the last chapter. The counties which are located in South Carolina and Georgia are classified as Old South; those along the river valleys of Louisiana, Mississippi, Arkansas and northern Alabama as Alluvial; those in eastern Texas and the Alabama "black belt" as New South. The sample contains 309 farms from the Old South, 365 from the Alluvial, and 223 farms from the New South.

TESTS OF THE REGIONAL SAMPLES

In Chapter III we tested the 897-farm sample selected from the 1850 manuscript census and concluded that there was insufficient evidence to warrant discarding the sample as being unrepresentative of the seventy-four county universe. Before examining the role of the small farmer in each of the regions in 1850, we should determine whether the regional subsamples appear to be representative of their respective universes. Due to the paucity of published data we are able to apply only one test to the 1850 sample. This test is the by now familiar comparison of the mean crop outputs per sample farm with the mean crop outputs per universe farm. The results of this test are summarized in Table 34. If a five percent level of significance is used, the only category in which the sample mean is significantly different from the universe mean is the improved acreage category in the Alluvial sample. Considering that there is a total of ten t-statistics per region, it is not surprising that one of them is significant at the five percent level. If we allow ourselves the liberty of considering the number of positive and negative t-statistics in each of the regions, we find that four of ten are negative in the Old South, seven of ten are negative in the Alluvial and eight of ten are negative in the New South. In the New South region a sign test is significant at the ten percent level but not at a five percent level.

TABLE 34

TESTS OF THE 1850 SAMPLE BY REGION
OLD SOUTH VS ALLUVIAL VS NEW SOUTH
(Acreage, Value of Farm, Crop Outputs)

	Old South Regions	Alluvial Regions	New South Regions
Number of farms in sample:	309	365	223
Values of t-statistics:			
Improved acres	.48	-2.27	.25
Unimproved acres	-.76	-.08	-.30
Value of farm	.73	.15	-.39
Value of farm machinery	.91	-.35	-.78
Swine	.77	-1.41	-1.31
Corn	-.36	-1.49	-1.07
Rice	-1.17	.52	.59
Tobacco	.94	.47	-.23
Cotton	.87	-.38	-.35
Sweet potatoes	-.77	-1.55	-.10

Source: Calculated from 1850 manuscript census sample and U.S. Bureau of the Census, Seventh Census of the United States: 1850, Compendium.

The results of this one test indicate that there is insufficient reason to reject the regional samples as being unrepresentative of their universes. However, we should note that the New South sample and possibly the Alluvial sample may be biased in the direction of farms which are smaller than those in the universe. (Again, this bias is exactly opposite to the one which we would have preferred for the purposes of this paper.)

INTERREGIONAL COMPARISONS

The size distributions of sample farms in each of the three regions are compared in Table 35. Smaller farms were relatively more numerous in the New South region than in either of the other regions. In the Alluvial region small farms were more common than in the Old South but not nearly so predominant as in the New South. In fact, using a .10 level of significance the distribution of farm sizes in the Alluvial region was not significantly different from that of either the Old or New South although the distribution clearly occupied a middle ground position between the Old South and New South extremes. In the New South region approximately 9 percent of the farms contained less than twenty improved acres and 43 percent contained less than fifty. In the Old South these percentages were 6 percent and 27 percent respectively. If farms of less than one hundred acres are compared the results are the same--New South, 65 percent; Old South, 53 percent. The distributions of farm sizes in the New and Old South regions were significantly different at the .05 level.

As Table 36 indicates, nonslaveholders were relatively more numerous in the New South than in either of the other regions. Nearly one-half of the farmers in the New South region did not own any slaves as compared with one-fourth to one-third of the farmers in the other regions. The size distribution of slaveholdings in each of the regions were significantly different at the .001 level. If we exclude nonslaveholders from the distributions, much of the difference between the New South and Alluvial distributions and the Old South and New South distributions disappears. (See Table 37.) However, if nonslaveholders are omitted, the New South distribution occupied a middle position and the Old South

TABLE 35

SIZE DISTRIBUTION OF FARMS (IMPROVED ACREAGE)
OLD SOUTH VS ALLUVIAL VS NEW SOUTH--1850

Farm Sizes	Old South		Alluvial		New South	
Less than 3 acres	1	(0.3%)	3	(0.8%)	1	(0.4%)
3-9 acres	4	(1.3%)	8	(2.2%)	3	(1.3%)
10-19 acres	14	(4.5%)	29	(7.9%)	16	(7.2%)
20-49 acres	64	(20.7%)	89	(24.4%)	76	(34.1%)
50-99 acres	79	(25.6%)	70	(19.2%)	49	(22.0%)
100-499 acres	126	(40.8%)	133	(36.4%)	65	(29.1%)
500-999 acres	17	(5.5%)	28	(7.7%)	9	(4.0%)
1000 acres & larger	4	(1.3%)	5	(1.4%)	4	(1.8%)
Total	309		365		223	

$x^2(df=7) = 10.52$
$p = .17$

$x^2(df=7) = 11.31$
$p = .13$

$x^2(df=7) = 16.82$
$p = .02$

Source: Calculated from 1850 manuscript census sample.

TABLE 36

SIZE DISTRIBUTION OF SLAVEHOLDINGS
OLD SOUTH VS ALLUVIAL VS NEW SOUTH--1850

Slaveholdings	Old South	Alluvial	New South
0 slaves	86 (27.8%)	107 (29.3%)	102 (45.7%)
1-4 slaves	77 (24.9%)	54 (14.8%)	40 (17.9%)
5-9 slaves	59 (19.1%)	45 (12.3%)	21 (9.4%)
10-19 slaves	45 (14.6%)	65 (17.8%)	30 (13.5%)
20-49 slaves	33 (10.7%)	62 (17.0%)	17 (7.6%)
50-99 slaves	6 (1.9%)	25 (6.8%)	12 (5.4%)
100-199 slaves	3 (1.0%)	6 (1.6%)	1 (0.4%)
200-499 slaves	---	1 (0.3%)	---
500-999 slaves	---	---	---
1000 slaves & over	---	---	---
Total	309	365	223

$x^2(df=6) = 29.58$, $p < .0005$ $x^2(df=6) = 25.81$, $p < .0005$

$x^2(df=6) = 29.15$, $p < .0005$

Source: Calculated from 1860 manuscript census sample.

TABLE 37

SIZE DISTRIBUTION OF SLAVEHOLDINGS (EXCLUDING NONSLAVEHOLDERS)
OLD SOUTH VS ALLUVIAL VS NEW SOUTH

Slaveholdings		Old South		Alluvial		New South	
1-4	slaves	77	(34.5%)	54	(20.9%)	40	(33.1%)
5-9	slaves	59	(26.5%)	45	(17.4%)	21	(17.4%)
10-19	slaves	45	(20.2%)	65	(25.2%)	30	(24.8%)
20-49	slaves	33	(14.8%)	62	(24.0%)	17	(14.0%)
50-99	slaves	6	(2.7%)	25	(9.7%)	12	(9.9%)
100-199	slaves	3	(1.3%)	6	(2.3%)	1	(0.8%)
200-499	slaves	---		1	(0.4%)	---	
500-999	slaves	---		---		---	
1000 slaves & over		---		---		---	
Total		223		258		121	

$x^2(df=5) = 29.40$
$p < .0005$

$x^2(df=5) = 10.09$
$p = .08$

$x^2(df=5) = 11.68$
$p = .04$

Source: Calculated from 1860 manuscript census sample.

and Alluvial distributions remain significantly different at the .001 level. The difference in the Old South and Alluvial distributions was due to the relative predominance of large slaveholders in the Alluvial regions. It is interesting to note that in both the Alluvial and New South regions over ten percent of the slaveholders owned fifty slaves or more. In the Old South only four percent of the slaveholdings were that large.

Thus, we find that small farms made up a greater proportion of farms in the New South than in the Old South. This situation obtains whether one considers farms of less than three acres, three to nine acres, ten to nineteen acres, or twenty to forty-nine acres of improved land. The size distribution of farms in the Alluvial region was between those of the New and Old South. Nonslaveholdings were much more common in the New South region than in either of the other regions. Finally, the size distributions of slaveholdings (excluding nonslaveholders) in the New and Old South were significantly different. Further, larger slaveholdings were relatively more common in the Allivial than in the other two regions. Again, the hypothesis that large planters led the migration into the Southwest appears questionable.

Having determined that there were relatively more small farmers and nonslaveholders in the New South than in the Old South in 1850, we now turn to a consideration of the relative roles played by the smaller farmers in each of the regions. Crop outputs, wealth, acreage and population data by acreage class expressed as a percentage of regional totals are presented in Table 38. In every category except rice, cotton and sweet potato production, the share of the total New South property owned or crops produced by farmers with less than twenty improved acres exceeds

TABLE 38

CROP OUTPUT, WEALTH, ACREAGE, AND POPULATION BY ACREAGE CLASSES
AS A PERCENT OF REGIONAL TOTALS
OLD SOUTH VS ALLUVIAL VS NEW SOUTH - 1850

		Farm Sizes (Improved Acreage)					
Item	Region	Less than 20	20-49	50-99	100-499	500-999	1000 & Larger
No. of Farms	OS	6.1%	20.7%	25.6%	40.8%	5.5%	1.3%
	All	10.9	24.4	19.2	36.4	7.7	1.4
	NS	8.9	34.1	22.0	29.1	4.0	1.8
Val. of Real Estate	OS	1.0	5.5	9.4	53.1	21.1	9.9
	All	.9	3.2	7.3	38.0	29.8	20.8
	NS	1.1	6.8	7.9	42.8	28.3	13.0
Improved Acreage	OS	.4	4.3	10.6	49.1	22.2	13.3
	All	.7	4.4	7.7	45.8	31.3	10.0
	NS	.9	8.3	11.1	43.8	22.0	14.0
Unimproved Acreage	OS	.7	8.2	13.1	48.9	15.0	14.1
	All	1.8	6.1	10.1	40.9	27.7	13.3
	NS	2.9	21.0	9.6	47.4	10.8	8.3
Val. of Farm	OS	.7	4.9	9.4	49.6	20.9	14.5
	All	.9	3.2	5.7	35.7	35.9	18.6
	NS	1.1	6.2	7.8	39.0	29.8	16.0
Val. of Farm Mach.	OS	1.0	5.0	9.1	53.7	19.1	12.0
	All	.9	2.9	7.1	45.5	33.8	9.7
	NS	2.7	7.1	10.7	47.4	19.8	12.3
Swine	OS	1.9	8.7	15.6	50.3	15.2	8.4
	All	4.1	13.3	17.9	43.7	17.0	4.0
	NS	3.5	19.8	16.4	37.9	14.3	8.1

TABLE 38 (Continued)

Item	Region	Farm Sizes (Improved Acreage)					
		Less than 20	20-49	50-99	100-499	500-999	1000 & Larger
Corn	OS	1.2%	6.3%	12.1%	52.9%	18.5%	9.0%
	All	2.3	8.3	11.1	44.0	24.5	9.8
	NS	1.9	11.4	12.5	40.2	24.0	9.9
Rice	OS	7.1	10.9	5.1	68.6	1.2	7.1
	All	---	30.7	18.5	23.6	---	27.2
	NS	---	4.6	11.1	53.8	30.5	---
Tobacco	OS	---	---	1.8	98.2	---	---
	All	5.7	28.1	45.4	20.7	---	---
	NS	---	85.7	14.3	---	---	---
Cotton	OS	.9	2.9	8.4	46.9	26.7	14.2
	All	.3	2.4	4.9	41.2	38.6	12.6
	NS	.7	4.2	8.0	38.0	35.1	14.0
Sweet Potatoes	OS	3.6	8.0	13.5	51.8	16.1	6.9
	All	2.1	8.8	12.0	53.5	19.9	3.7
	NS	3.1	12.8	13.1	46.5	17.1	7.3
Free Males	OS	4.3	16.4	28.5	43.0	6.5	1.2
	All	9.0	23.2	20.2	39.7	6.6	1.3
	NS	6.6	34.8	25.9	26.9	3.5	2.3
Free Females	OS	5.3	19.3	26.9	42.1	5.2	1.1
	All	9.0	27.8	20.9	34.7	6.4	1.0
	NS	7.9	35.4	21.6	29.7	3.0	2.6
Slave Males	OS	1.1	1.6	6.4	51.5	25.9	13.4
	All	.8	2.5	5.8	47.2	30.9	12.9
	NS	.7	3.1	6.2	46.3	29.4	14.3

TABLE 38 (Continued)

Item	Region	Farm Sizes (Improved Acreage) Less than 20	20-49	50-99	100-499	500-999	1000 & Larger
Slave Females	OS	1.6%	2.3%	7.6%	50.7%	24.6%	13.1%
	All	.7	3.7	6.2	46.4	30.8	12.2
	NS	.8	2.6	7.0	44.7	28.7	16.2
Total Population	OS	2.6	7.8	14.7	48.0	18.1	8.8
	All	2.8	8.9	9.8	44.3	24.4	9.6
	NS	3.4	16.1	13.6	38.5	18.4	9.9

Source: Calculated from 1850 manuscript census sample.

the share of property held or crops produced by their counterparts in the Old South. The percentage of free persons living on farms of less than twenty acres in the New South exceeded that in the Old South. However, the percentage of slaves held by farmers with less than twenty acres in the New South was less than that of their Old South counterparts. If we consider farms of less than fifty improved acres, we find that the share of total crops produced by the smaller farmers in the New South exceeded that of smaller farmers in the Old South in every category except rice production. Over forty percent of the free population (rural) were found on farms of less than fifty improved acres in the New South as compared to less than twenty-five percent in the Old South.

The percentage of Alluvial property owned or output produced on farms of less than twenty acres were approximately the same as in the Old South region for most categories. The same observation obtains if farms of less than fifty improved acres are compared. However, an exception was cotton production. In the Alluvial region, it appears that cotton production was concentrated on larger farms to a greater extent than was the case in either of the other regions. Also, it is interesting to note that, although the share of crop outputs of smaller farmers in the Alluvial region were generally less than that of their New South counterparts, small farmers in the Alluvial region led in the percentage of swine, corn and rice produced.

Crop outputs, wealth, acreage and population data by slaveholding class are expressed as a percentage of regional totals in Table 39. Nonslaveholders in the New South accounted for a greater share of property owned and crops produced than did their counterparts in the Old South in every category except rice production. The same observation obtains if

TABLE 39

CROP OUTPUT, WEALTH, ACREAGE AND POPULATION BY SLAVEHOLDING CLASSES
AS A PERCENT OF REGIONAL TOTALS
OLD SOUTH VS ALLUVIAL VS NEW SOUTH - 1850

		Size of Slaveholding							
Item	Region	0	1-4	5-9	10-19	20-49	50-99	100-199	200 & Larger[a]
No. of Farms	OS	27.8	24.9	19.1	14.6	10.7	1.9	1.0	---
	All	29.3	14.8	12.3	17.8	17.0	6.8	1.6	.3
	NS	45.7	17.9	9.4	13.5	7.6	5.4	.4	---
Val. of Real Estate	OS	5.6	12.4	13.4	22.6	34.0	5.0	6.9	---
	All	3.0	4.6	4.1	11.0	26.2	24.0	10.7	16.4
	NS	9.3	8.5	6.2	17.2	18.1	32.3	8.3	---
Improved Acreage	OS	8.2	11.3	12.8	19.4	29.6	8.9	9.7	---
	All	7.9	5.0	5.4	15.4	28.0	27.4	7.7	3.3
	NS	13.5	9.3	6.7	18.4	16.9	32.1	3.1	---
Unimproved Acreage	OS	9.6	12.1	13.4	15.6	29.4	3.9	16.0	---
	All	7.6	5.8	6.5	15.2	24.2	22.0	7.4	11.3
	NS	22.2	16.4	7.0	23.9	10.7	16.6	3.1	---
Val. of Farm	OS	6.0	10.8	11.3	20.5	27.7	10.4	13.3	---
	All	3.9	3.1	3.5	10.3	25.0	26.6	13.3	14.4
	NS	9.5	8.0	5.7	16.2	15.3	37.6	7.7	---
Val. of Farm Mach.	OS	5.0	10.5	14.4	26.3	22.8	11.0	10.1	---
	All	5.3	2.8	4.3	14.3	30.2	27.4	10.4	5.2
	NS	11.1	9.2	8.0	20.7	17.3	31.0	2.7	---
Swine	OS	12.1	15.0	16.3	20.1	22.5	6.3	7.6	---
	All	16.5	8.3	10.3	18.0	27.4	14.9	4.0	.7
	NS	24.9	13.1	7.5	18.8	13.8	20.8	1.0	---

TABLE 39 (Continued)

Item	Region	Size of Slaveholding							
		0	1-4	5-9	10-19	20-49	50-99	100-199	200 & Larger[a]
Corn	OS	9.3	11.9	15.5	20.0	26.7	8.0	8.6	---
	All	11.2	5.6	8.4	15.0	26.4	23.0	6.1	4.3
	NS	14.8	10.7	8.4	17.9	16.3	29.1	2.9	---
Rice	OS	20.4	1.4	.1	7.1	63.8	---	7.1	---
	All	10.2	21.8	9.7	25.3	5.8	---	27.3	---
	NS	1.7	10.8	5.6	40.6	10.8	30.5	---	---
Tobacco	OS	---	---	---	28.6	71.4	---	---	---
	All	50.9	13.6	15.8	19.8	---	---	---	---
	NS	85.7	14.3	---	---	---	---	---	---
Cotton	OS	4.7	7.2	12.0	19.0	28.5	13.2	15.4	---
	All	3.0	2.4	3.1	13.2	28.3	33.6	11.2	5.3
	NS	5.9	7.4	6.5	15.7	17.7	37.9	8.8	---
Sweet Potatoes	OS	12.1	12.1	16.7	24.5	20.8	5.5	8.3	---
	All	9.5	8.1	8.5	22.3	27.3	19.7	2.0	2.7
	NS	15.7	9.6	9.2	27.4	13.6	21.7	2.7	---
Free Males	OS	26.1	25.2	19.2	14.4	12.2	2.2	.6	---
	All	29.4	15.2	10.6	19.9	16.7	6.3	1.6	.2
	NS	48.8	19.2	8.4	11.4	6.4	5.0	.8	---
Free Females	OS	29.8	25.0	19.3	12.6	10.9	1.8	.6	---
	All	32.5	15.9	12.5	15.9	16.0	5.8	1.4	.1
	NS	49.0	17.3	9.2	11.9	6.5	5.7	.3	---
Slave Males	OS	---	4.2	12.8	21.4	35.2	13.4	13.1	---
	All	---	1.8	5.1	14.6	32.3	28.8	13.0	4.5
	NS	---	4.3	6.3	21.0	22.5	39.7	6.2	---

TABLE 39 (Continued)

Item	Region	Size of Slaveholding							
		0	1-4	5-9	10-19	20-49	50-99	100-199	200 & Larger[a]
Slave Females	OS	---	6.0	14.8	20.5	31.7	13.4	13.7	---
	All	---	2.2	5.3	15.0	31.6	29.5	12.4	3.8
	NS	---	4.1	7.3	20.3	22.3	38.7	7.3	---
Total Population	OS	10.2	12.4	15.8	18.2	25.4	9.2	8.7	---
	All	8.1	5.6	6.8	15.7	27.9	23.1	9.8	3.1
	NS	20.1	10.0	7.6	17.0	15.9	25.3	4.2	---

[a]The sample contained only one farm having more than 200 slaves.

Source: Calculated from 1850 manuscript census sample.

one considers farmers with less than five slaves (including nonslaveholders). The percentages of wealth owned and crops produced by nonslaveholders and small slaveholders in the Alluvial region were generally much lower than these percentages in each of the other regions. Evidently, wealth holdings and crop production were concentrated in the larger slaveholdings to a greater degree in the Alluvial region than in the other two regions.

CONCLUSIONS

Our analysis of the 1850 sample yields four interesting results. First, small farms and nonslaveholders were more common in the New South than in the Old South region. Nonslaveholders were relatively more numerous in the New South than in the Alluvial region but the size distributions of farms in the two regions were substantially the same.

Second, if nonslaveholders are omitted from the distributions, the size distributions of slaveholdings in the New and Old South were significantly different at the five percent level. Larger slaveholdings were slightly more common in the New South. In the Alluvial region, larger slaveholdings were even more common. (The Alluvial size distribution of slaveholdings was significantly different from that of the Old South at the .001 level.)

Third, small farmers and small slaveholders in the New South accounted for a greater share of wealth owned and crops produced than did their counterparts in the Old South. In the Alluvial region small farmers produced approximately the same share of regional crop production as did their counterparts in the Old South but small <u>slaveholders</u> produced a considerably smaller share of regional crop production in the Alluvial than in either the Old or New South.

Finally, a much greater percentage of the rural free population in the New South lived on farms without slaves than in either the Alluvial or Old South regions. Also, a greater percentage of the rural free population in the Alluvial and New South regions lived on smaller farms than in the Old South.

These findings suggest that small farmers and nonslaveholders were certainly more common in the newer regions (New South and Alluvial) than in the older regions in 1850. However, in the Alluvial region, it appears that, while farms may have been smaller than in the Old South, slaveholdings were not smaller but larger. And, in the New South region, while farms were smaller and nonslaveholders relatively more numerous than in the Old South, when slaves *were* owned, the holdings tended to be larger than in the Old South.

Clearly, the 1850 data fail to support our hypothesis that large planters led the migration into the new lands in the Southwest. However, the tendency toward smaller farms but larger slaveholdings in the Alluvial region (as compared to the Old South) may suggest that large planters either led the way into the Alluvial or pushed the smaller slaveholders out. (An alternative conclusion would be that the small slaveholders prospered.) As in 1860, the tendency toward smaller farms might be explained by the greater fertility of the Alluvial soils. Presumably, if the soil of a region is more fertile, the size of a farming operation would be limited by the number of slaves which could be purchased and managed rather than the number of acres which could be cultivated. Thus, in regions of rich soil one would expect to find relatively smaller farms with relatively large slaveholdings and, in poorer regions, just the reverse.

CHAPTER VII

THE DISPLACEMENT OF THE SMALL FARMER

In the two previous chapters the first part of our hypothesis has taken a beating. In fact, it is the writer's opinion that the beating was fatal. Next, we consider the second part of the hypothesis in an attempt to salvage some part of the traditional argument. As it turns out, a part of the hypothesis is salvageable: the small farmer was being displaced. But, he was not being displaced in the New South. Once again, we are ahead of our story.

Before turning to a comparison of the 1850 and 1860 data perhaps we should restate the second part of the hypothesis presented in Chapter I:

> "In the case that the large planters immediately followed the smaller farmer into the newer regions, the smaller farmers were pushed out by larger planters who were able to buy the better land from them."

THE REGIONS CONSIDERED

If it is true that the large planters moved into the newer regions and bought out the smaller farmers, one would expect to find that large planters were more common in the newer regions in 1860 than in 1850. Strictly speaking, we would expect to find large planters becoming more predominant only in the "better" regions. Presumably, however, "better land" exists in most of the major regions where cotton was extensively cultivated although it is generally considered that the alluvial land along the larger rivers (Mississippi, Yazoo, Red, and Tennessee Rivers) was the most favorable land for cotton production. Thus, although we would expect to find a greater percentage of large planters in the Southwest in 1860 than in 1850, we would expect the difference to be more pronounced in the Alluvial region than in the other parts of the New South. In the remainder of this chapter, the 1850 and 1860 samples are compared in an attempt to determine the changes in size of farms and size of slaveholdings which took place during the decade between the 1850 and 1860 censuses.

Before proceeding to a comparison of the 1850 and 1860 data we should define the regions being considered. It should be recalled that the entire 5229 farm 1860 manuscript census sample is selected from the counties which produced over 1000 bales of cotton in 1859. The 1850 manuscript census sample is selected only from the seventy-four counties described by Gray as the major cotton producing counties in 1859. (Actually two of the seventy-four counties were not counties in 1850 and the manuscript census data for five others have been lost. Consequently, farms in only sixty-seven counties actually had a chance to be included in the sample. The seven counties which were omitted from the 1850 universe are located in the Old South region.) To insure comparability between the 1850 and 1860 samples, the subsample of the 1860 sample which consists of 964 farms selected from Gray's seventy-four major cotton producing counties is considered in this chapter.

Gray divides the seventy-four county region into the three subregions--Old South, Alluvial and New South--which were used in the analyses presented in Chapters IV and V. These three regions are again utilized in this chapter.

At this point, we should review the apparent biases in each of the regional samples. Briefly the biases are:

Old South--1850: No apparent bias.

Old South--1860: Biased in the direction of too many large farms.

Alluvial--1850: Possibly biased in the direction of too many small farms.

Alluvial--1860: No apparent bias.

New South--1850: Very possibly biased in the direction of too many small farms. (This conclusion is based on a comparison of mean crop output per sample farm with that of the universe farms. Consequently the evidence is, at best, indirect.)

New South--1860: Biased in the direction of too many small farms.

These biases must be borne in mind when the 1850 and 1860 data are compared. Fortunately, for the purposes of this chapter, only one of these biases presents any major problem. As we shall later observe, there appears to have been a greater percentage of large farms in the Old South in 1860 than in 1850. Because the 1860 Old South sample is biased in the direction of too many large farms, it is conceivable that the greater percentage of large farms in 1860 than in 1850 could be the result of this bias. Fortunately, data is available to demonstrate that the 1850-1860 differences are not merely the results of sample bias. The Alluvial samples are both quite good. However, the possibility of a bias in the direction of too many small farms in the 1850 sample, requires that we qualify our findings. Finally, the biases in the New South in both 1850 and 1860 are in the direction of too many small farms. Because these biases are in the same direction in both years, comparisons between the 1850 and 1860 data probably can be made without qualification or adjustment. However, by using additional data to remove the bias in the 1860 New South sample, we can greatly strengthen the implication of our results. These modifications are considered in detail later in the chapter.

One further source of error should be elaborated on at this point. As mentioned above, it was impossible to sample seven of the universe counties in 1850. As all seven of these counties were located in the Old South region, the 1850 and 1860 Old South samples are actually selected from different regions (different by seven counties). Fortunately, it is possible to utilize data from the published census to demonstrate that the differences in the counties sampled are not sufficient to influence the 1850-1860 comparison significantly.

COMPARISON OF THE 1850 AND 1860 DATA

In Table 40, the size distribution of farms in each of the three regions in the 1850 and 1860 census years are contrasted. To make the comparison easier, the number of sample farms in each improved acreage category in each census year is expressed as a percentage of the total sample farms in that particular region and year. (There are 309 Old South farms in the 1850 manuscript census sample. One farm contained less than three improved acres. Thus, the first entry in Table 40 is: 1 + 309 = .0032 or 0.3%.)

TABLE 40

SIZE DISTRIBUTION OF FARMS (IMPROVED ACREAGE)
1850 VS 1860

Farm Sizes	Old South 1850	Old South 1860	Alluvial 1850	Alluvial 1860	New South 1850	New South 1860
Less than 3 acres	0.3%	1.8%	0.8%	0.5%	0.4%	1.6%
3-9 acres	1.3%	0.7%	2.2%	1.9%	1.3%	3.0%
10-19 acres	4.5%	2.2%	7.9%	7.7%	7.2%	10.5%
20-49 acres	20.7%	14.8%	24.4%	16.3%	34.1%	25.0%
50-99 acres	25.6%	18.8%	19.2%	18.7%	22.0%	23.0%
100-499 acres	40.8%	46.3%	36.4%	38.0%	29.1%	33.2%
500-999 acres	5.5%	8.8%	7.7%	12.5%	4.0%	2.3%
1000 acres & larger	1.3%	6.4%	1.4%	4.3%	1.8%	1.3%
	x^2(df=7) = 30.10 $p < .0005$		x^2(df=7) = 13.53 $p = .06$		x^2(df=7) = 9.08 $p = .25$	

Source: Calculated from the 1850 and 1860 manuscript census samples.

Quite frequently, distributions similar to those in Table 40 are compared and "obvious" differences are noted without any attempt to determine whether these differences are significant or merely a chance occurrence. A chi-square statistic may be employed to determine the probability that the differences between two distributions are merely the result of chance. This statistic, calculated from the actual distributions rather than the percentages, appears at the bottom of each pair of columns in Table 40. The probability level calculated from this statistic indicates the probability that the differences in the 1850 and 1860 distributions arise merely as a result of chance, that is, the probability that the 1850 and 1860 distributions are not really different and the apparent differences are merely the result of the fact that two samples rather than two populations are being compared.

First, we should note that the only census-to-census difference which is significant at the .05 level occurs in the Old South region. Clearly, the data presented in Table 40 indicates that between 1850 and 1860 there was a significant increase in the percentage of farms in the larger acreage classes and, of course, a concomitant decline in the percentage of farms in the smaller improved acreage classes. Although the percentage of farms having less than three acres increased between 1850 and 1860, the percentage in all classes from three acres to ninety-nine acres declined. Thus, in 1850, 52.4 percent of the Old South farms were smaller than one hundred acres and only 47.6 larger. In 1860, only 38.3 percent were smaller than one hundred acres.

In the Alluvial region, the differences between the 1850 and 1860 distributions are not significant at the .05 level. However, the probability level is approximately six percent and, consequently, it is not completely unreasonable to examine the distributions for differences which might prove interesting. As in the Old South, the percentage of Alluvial farms in the larger acreage classes increased between 1850 and 1860. In 1850, only 45.4 percent of the Alluvial farms contained one hundred or more improved acres. By 1860, 54.8 of the Alluvial farms were this large. Thus, although the 1850 and 1860 size distributions of Alluvial farms are not significantly different, there is some indication that the percentage of larger farms in the Alluvial region increased between 1850 and 1860.

The size distributions of farms in the New South region in 1850 and 1860 are not significantly different. However, one is always tempted to point out the differences, e.g., the increase in the percentage of farms of less than twenty acres and one hundred acres or larger. Clearly, this is not appropriate in view of the .25 probability level. Thus, we take the position that the 1850 and 1860 New South distributions were not significantly different.

Now, what are the implications of the above results? Most important, there is absolutely no indication that the large planter was displacing the smaller farmer in the New South region--a region, incidentally, which includes the fertile "black belt" of Alabama. Second, large planters <u>may</u> have been displacing the smaller farmers in the rich Alluvial lands. We say "may" due to the .06 probability level and the corollary danger of bending our definition of significance to suit the task at hand. However, the fact that smaller planters in the Alluvial region may have been displaced provides us with a clue to the source of our hypothesis. Once again, the Alluvial region appears to have been overemphasized. Third, the percentage of large farms in the Old South region increased significantly between 1850 and 1860, a time when the number of farms in the region was decreasing. Clearly, if smaller farmers were being bought out in any region, it was in the Old South. Thus, the results do not support the second part of our hypothesis. Finally, this comparison of the size distribution of farms in the three regions in 1850 and 1860 casts even more doubt on the validity of the first part of the hypothesis. Clearly, the increase in the percentage of large farms in the Old South combined with no significant change in the New South is exactly the opposite finding to what one would expect if the hypothesis were correct. The larger planters appear to have bought out smaller farmers in the Old South thereby both encouraging the smaller farmer to move west and planting their own roots still deeper in the East.

At this point the reader likely recalls the biases mentioned earlier and is wondering whether the above findings might merely result from the deficiencies of the samples. The answer is no--the results are not the consequence of errors in the sample. The effects of the biases on each of the regions are discussed below.

The 1850-1860 comparison of the Old South sample data can be questioned on two grounds. First, the 1850 and 1860 data are selected from different universes. Seven of the Old South counties included in the 1860 sample could not be included in the 1850 sample. Second, while the 1850 sample appears to be unbiased, the 1860 sample is clearly biased in the direction of too many large farms. Fortunately, the effects of each of these factors are easily evaluated using data from the published census. (Although the published census for 1850 does not contain a size distribution of farms by county, this data is available for 1860. Fortunately, the 1860 data will suffice.) In Table 41, the size distribution of Old South farms in the 1850 <u>sample</u> is compared to the size distribution of Old South farms calculated from the <u>published census</u> for 1860. The 1850 sample appears unbiased and the 1860 universe data (from the published census) is certainly unbiased. Further, the 1860 data is calculated using <u>only</u> the counties from which the 1850 sample is selected. Thus, bias is removed and comparability insured. Using this new comparison (Table 41), we still find an increase in the percentage of larger farms in the Old South region between 1850 and 1860. In 1850, 47.7 percent of the Old South farms with three or more acres contained

contained one hundred improved acres or more. By 1860, this percentage had increased to 58.8. A comparison of Table 41 with Table 40 indicates that the 1850-1860 differences in the size distributions of Old South farms are not merely the result of sample bias.

TABLE 41

SIZE DISTRIBUTION OF FARMS (IMPROVED ACREAGE)
1850 VS 1860 (ADJUSTED DATA)

Farm Sizes	Old South 1850	1860	New South 1850	1860
3-9 acres	1.3	.7	1.3	2.6
10-19 acres	4.5	2.6	7.2	7.4
20-49 acres	20.8	16.1	34.3	28.3
50-99 acres	25.7	21.8	22.1	22.4
100-499 acres	40.9	47.7	29.2	31.6
500-999 acres	5.5	8.2	4.0	5.6
1000 acres & larger	1.3	2.9	1.8	2.3

Source: 1850 - Calculated from the manuscript census sample.
1860 - Calculated from U.S. Bureau of the Census, _Eighth Census of the United States: 1860, Agriculture_.

Both the 1850 and 1860 Alluvial samples appear to be representative of their respective universes. However, since the 1850 Alluvial sample may possibly be biased in the direction of too many small farms, we must be careful not to overemphasize the decline in the percentage of small farmers in the region between 1850 and 1860. Clearly, a part of the apparent shift may stem from sample bias. Although it is the writer's opinion that the effect of this bias is slight, the effect of the bias is to overstate the differences in the 1850 and 1860 distributions.

Removal of the bias, were it possible, would merely cast further doubt on the validity of the hypothesis.

In the New South, both the 1850 and 1860 samples are biased in the direction of too many small farms. Because distributions of farm sizes in individual counties were not published for 1850, the 1850 bias cannot be removed. The effect of the 1860 bias can be evaluated using the universe data calculated from the published census. A comparison of the 1850 sample data with the 1860 universe data for the New South is included in Table 41. As in the case of the Old South, it is clear that the 1860 sample bias has little effect. In fact, Table 41 overstates the differences in the regions. If the bias in the 1850 sample were removed the distributions would be even more similar.

Thus, the increase in the percentage of larger farms in the Old South and the lack of a significant change in the size distribution of farms in the New South are not the result of biases in the samples. Our observation that the farm-size data tend to contradict the second part of our hypothesis remains intact.

In Table 42, the size distributions of slaveholdings in each of the three regions in the 1850 and 1860 census years are compared. Again, the number of sample farms in each slaveholding category in each census year is expressed as a percentage of the total number of sample farms in that particular region and year. Also, a chi-square statistic, calculated from the actual distributions rather than from the percentages, is used to determine the probability that the differences between the 1850 and 1860 distributions are the result of chance. This statistic appears at the bottom of each pair of columns in Table 42.

TABLE 42

SIZE DISTRIBUTION OF SLAVEHOLDINGS
1850 VS 1860

Slaveholdings		Old South 1850	Old South 1860	Alluvial 1850	Alluvial 1860	New South 1850	New South 1860
0	slaves	27.8%	29.2%	29.3%	26.9%	45.7%	49.0%
1-4	slaves	24.9%	13.3%	14.8%	12.1%	17.9%	13.2%
5-9	slaves	19.1%	16.7%	12.3%	13.0%	9.4%	12.5%
10-19	slaves	14.6%	17.5%	17.8%	13.4%	13.5%	12.8%
20-49	slaves	10.7%	15.1%	17.0%	21.6%	7.6%	9.9%
50-99	slaves	1.9%	6.9%	6.8%	7.6%	5.4%	2.0%
100-199	slaves	1.0%	1.3%	1.6%	4.8%	0.4%	.7%
200-499	slaves	---	---	0.3%	0.5%	---	---
500-999	slaves	---	---	---	---	---	---
1000 slaves & over		---	---	---	---	---	---
		$x^2(df=6) = 27.41$ $p < .0005$		$x^2(df=6) = 9.17$ $p = .17$		$x^2(df=6) = 8.62$ $p = .20$	

Source: Calculated from the 1850 and 1860 manuscript census samples.

Once again, the only census-to-census difference which is significant at the .05 level (in fact, the only one at the .15 level) occurs in the Old South region. In this region, there appears to have been a significant increase in the percentage of large slaveholdings between 1850 and 1860. In 1850, only 28.2 percent of the farmers held ten or more slaves. By 1860, this percentage increased to 40.8. Surprisingly, the percentage of nonslaveholders increased slightly (but probably not significantly).

In the Alluvial and New South regions, the 1850 and 1860 size distributions of slaveholdings were not significantly different. Consequently, any observed differences are not to be taken too seriously and the following discussion of the 1850-1860 differences is offered with the caution that the differences are, statistically speaking, nonexistent. In the Alluvial region, there appears to have been an increase in the percentage of slaveholdings of twenty slaves and larger. In 1850, 25.7 percent of the farmers held twenty slaves or more. By 1860, this figure increased to 34.5 percent. The percentage of non-slaveholders in the region declined from 29.3 percent to 26.9 percent during the decade. In the New South, the 1850-1860 differences were somewhat the opposite. In 1850, 13.4 percent of the New South farmers owned twenty slaves or more and 26.9 percent owned ten or more slaves. By 1860, these figures had declined to 12.6 percent and 25.4 percent respectively. During the period, the percentage of New South nonslaveholders increased from 45.7 percent to 49.0 percent. Again, these differences should not be considered to be significant. However, it should be noted that any tendencies which the data may reveal fail to support the hypothesis that the larger planters bought out the smaller farmers with the significant exception of the Alluvial region. In this region, it appears that the percentage of large slaveholders _may_ have increased between 1850 and 1860. Again, generalizations regarding southwestward migration appear to be rooted in Alluvial. Clearly, the observation that the percentage of slaveholders with ten or more slaves in the Old South region increased from 28.2 percent in 1850 to 40.8 percent in 1860 casts further doubt on the validity of our hypothesis.

Table 43 presents a comparison of the size distributions of slaveholding in the three regions in 1850 and 1860 if nonslaveholders are eliminated from the distribution. Once again, the data are presented in percentages; chi-square statistics, calculated from the original data, are presented.

TABLE 43

SIZE DISTRIBUTION OF SLAVEHOLDINGS (EXCLUDING NONSLAVEHOLDERS) 1850 VS 1860

Slaveholdings	Old South 1850	Old South 1860	Alluvial 1850	Alluvial 1860	New South 1850	New South 1860
1-4 slaves	34.5%	18.8%	20.9%	16.3%	33.1%	25.8%
5-9 slaves	26.5%	23.7%	17.4%	17.7%	17.4%	24.6%
10-19 slaves	20.2%	24.7%	25.2%	18.5%	24.8%	25.1%
20-49 slaves	14.8%	21.3%	24.0%	29.6%	14.0%	19.3%
50-99 slaves	2.7%	9.7%	9.7%	10.5%	9.9%	3.9%
100-199 slaves	1.3%	1.9%	2.3%	6.6%	0.8%	1.3%
200-499 slaves	---	---	0.4%	.7%	---	---
500-999 slaves	---	---	---	---	---	---
1000 slaves & over	---	---	---	---	---	---
	$x^2(df=5) = 27.08$ $p < .0005$		$x^2(df=5) = 8.71$ $p = .13$		$x^2(df=5) = 7.91$ $p = .17$	

Source: Calculated from the 1850 and 1860 manuscript census samples.

As in each of the earlier distributions, the only significant 1850-1860 differences occurred in the Old South region. The Alluvial and New South distributions in 1850 were not significantly different from those in 1860 even at the .10 level. In the Old South, there was an increase in the percentage of slaveholders with 100-199 slaves, 50-99 slaves, 20-49 slaves, and 10-19 slaves between 1850 and 1860. Taken together, there was an increase in the percentage of slaveholdings of ten or more slaves from 39.0 percent in 1850 to 57.6 percent in 1860. Of course, the percentage of slaveholders in each of the other categories decreased but the drop was precipitous in the 1-4 slave category. Almost thirty-five percent of the Old South slaveholders owned one to four slaves in 1850 but less than twenty percent of the 1860 slaveholdings were in this category.

Although the 1850-1860 differences were not significant in the Alluvial region, there appears to have been an increase in the percentage of large slaveholders between 1850 and 1860. In the Alluvial region, the percentage of slaveholders who owned twenty slaves or more increased from 36.4 percent in 1850 to 47.4 percent in 1860. In the New South region, there was no apparent systematic change in the size distribution of slaveholders between 1850 and 1860. The percentage of slaveholders with one to four slaves decreased but the percentage with five to nine slaves increased by the same number of percentage points. Also, the percentage of slaveholders in the 20-49 slave category increased by five percent while the percentage in the 50-99 slave category decreased by six percent.

Thus, we find that between the 1850 and 1860 census years the percentage of larger slaveholders increased significantly in the Old South. In the Alluvial, the percentage of large slaveholders may have increased but not enough to be statistically significant. In the New South, there appears to have been no systematic change in the size distribution of slaveholdings (excluding nonslaveholders) between 1850 and 1860. The increase in the percentage of larger slaveholders in the Old South and the lack of any significant change in the New South distribution cast further doubt on the validity of the hypothesis. However, the increase in the percentage of large slaveholders in the Alluvial, although not statistically significant, supports both the hypothesis and the earlier observation that a myopic look at the Alluvial region may be the source of our hypothesis.

A further test of the second part of our hypothesis is a comparison of the role of the smaller farmers in each of the regions in 1850 to their role in the respective regions in 1860. Crop output, wealth, acreage and population data by acreage classes are expressed as a percentage of the 1850 and 1860 regional totals in Table 44. In the Old South region, the percentage of crops produced and wealth owned by farmers with more than five hundred improved acres increased between 1850 and 1860. Also, the percentages of the free and slave population living on these farms increased between the two census years. In the Alluvial region the changes were the same as in the Old South. In the New South, however, the changes between 1850 and 1860 were just the opposite. The percentage of New South crops produced and wealth owned by farmers with one hundred or more improved acres decreased. (Cotton, tobacco and sweet potato production are notable exceptions. The percentage of total output of each of these crops produced on farms of 100 to 499 improved acres increased during the period. However, the percentage produced on farms of more than five hundred acres declined markedly.) In the New South, the percentages of the free and slave population living on farms with five hundred acres or more decreased but the percentage living on farms with 100 to 499 acres increased during the decade. Again, we find evidence that the small farmers were being displaced in the Old South and in the Alluvial region but find no evidence of their displacement in the New South region.

Crop output, wealth, acreage and population data by slaveholding classes are expressed as a percentage of the 1850 and 1860 regional

TABLE 44

CROP OUTPUT, WEALTH, ACREAGE, AND POPULATION BY ACREAGE CLASSES AS A PERCENT OF REGIONAL TOTALS --1850 VS. 1860

			Farm Size (Improved Acreage)					
Item	Region	Year	Less Than 20	20-49	50-99	100-499	500-999	1000 & Larger
No. of Farms	OS	1850	6.1%	20.7%	25.6%	40.8%	5.5%	1.3%
		1860	4.7	14.8	18.8	46.5	8.8	6.4
	Al	1850	10.9	24.4	19.2	36.4	7.7	1.4
		1860	10.1	16.3	18.8	38.0	12.5	4.3
	NS	1850	8.9	34.1	22.0	29.1	4.0	1.8
		1860	15.1	25.0	23.0	33.3	2.3	1.3
Improved Acreage	OS	1850	.4	4.3	10.6	49.1	22.2	13.3
		1860	.1	1.8	5.0	38.1	22.0	33.1
	Al	1850	.7	4.4	7.7	45.3	31.3	10.0
		1860	.5	1.9	5.0	34.3	32.8	25.4
	NS	1850	.9	8.3	11.1	43.8	22.0	14.0
		1860	1.3	6.4	12.2	53.8	12.4	13.8
Unimproved Acreage	OS	1850	.7	8.2	13.1	48.9	15.0	14.1
		1860	.4	4.3	7.6	41.4	24.2	22.1
	Al	1850	1.8	6.1	10.1	40.9	27.7	13.3
		1860	1.3	5.1	12.2	36.7	23.6	20.9
	NS	1850	2.9	21.0	9.6	47.4	10.8	8.3
		1860	4.7	9.7	22.2	53.4	4.3	5.7
Val. of Farm	OS	1850	.7	4.9	9.4	49.6	20.9	14.5
		1860	.3	3.0	6.7	43.1	21.5	25.5
	Al	1850	.9	3.2	5.7	35.7	35.9	18.6
		1860	.2	.8	3.1	28.7	33.7	33.4
	NS	1850	1.1	6.2	7.8	39.0	29.8	16.0
		1860	2.1	5.9	12.3	56.6	12.6	10.8

TABLE 44 (Continued)

Item	Region	Year	Less Than 20	20-49	50-99	100-499	500-999	1000 & Larger
			Farm Size (Improved Acreage)					
Swine	OS	1850	1.9%	8.7%	15.6%	50.3%	15.2%	8.4%
		1860	1.1	5.8	9.4	41.9	19.7	22.3
	Al	1850	4.1	13.3	17.9	43.7	17.0	4.0
		1860	4.3	6.9	10.1	44.7	21.4	12.5
	NS	1850	3.5	19.8	16.4	37.9	14.3	8.1
		1860	4.9	16.2	16.1	52.8	7.6	2.3
Corn	OS	1850	1.2	6.3	12.1	52.9	18.5	9.0
		1860	.4	3.5	7.3	40.8	21.3	26.5
	Al	1850	2.3	8.3	11.1	44.0	24.5	9.8
		1860	1.3	2.6	6.3	38.7	31.9	19.1
	NS	1850	1.9	11.4	12.5	40.2	24.0	9.9
		1860	2.8	8.4	14.5	53.2	14.9	6.2
Tobacco	OS	1850	---	---	1.8	98.2	---	---
		1860	---	71.4	---	28.6	---	---
	Al	1850	5.7	28.1	45.4	20.7	---	---
		1860	---	---	---	---	---	---
	NS	1850	---	85.7	14.3	---	---	---
		1860	---	40.0	40.0	20.0	---	---
Cotton	OS	1850	.9	2.9	8.4	46.9	26.7	14.2
		1860	.1	1.7	4.8	34.8	22.6	35.9
	Al	1850	.3	2.4	4.9	41.2	38.6	12.6
		1860	.2	.8	3.6	34.2	33.5	27.6
	NS	1850	.7	4.2	8.0	38.0	35.1	14.0
		1860	.8	5.1	7.3	59.3	15.8	11.7

TABLE 44 (Continued)

Item	Region	Year	Less Than 20	20-49	50-99	100-499	500-999	1000 & Larger
			Farm Size (Improved Acreage)					
Sweet Potatoes	OS	1850	3.6%	8.0%	13.5%	51.8%	16.1%	6.9%
		1860	1.0	5.2	7.8	38.0	19.5	28.5
	Al	1850	2.1	8.8	12.0	53.5	19.9	3.7
		1860	2.3	5.7	9.3	48.2	23.0	11.5
	NS	1850	3.1	12.8	13.1	46.5	17.1	7.3
		1860	4.4	11.3	14.4	67.4	9.9	2.5
Free Males	OS	1850	4.3	16.4	28.5	43.0	6.5	1.2
		1860	3.5	13.9	19.2	48.9	8.6	6.0
	Al	1850	9.0	23.2	20.2	39.7	6.6	1.3
		1860	10.4	17.3	20.9	33.9	14.1	3.4
	NS	1850	6.6	34.8	25.9	26.9	3.5	2.3
		1860	10.7	23.5	26.2	35.6	3.1	.9
Free Females	OS	1850	5.3	19.3	26.9	42.1	5.2	1.1
		1860	3.7	15.2	18.8	45.9	8.6	7.8
	Al	1850	9.0	27.8	20.9	34.7	6.4	1.0
		1860	11.5	16.6	18.7	35.4	15.6	2.3
	NS	1850	7.9	35.4	21.6	29.7	3.0	2.6
		1860	13.5	24.7	25.1	33.2	2.7	.7
Slave Males	OS	1850	1.1	1.6	6.4	51.5	25.9	13.4
		1860	---	1.7	2.7	39.1	24.7	31.7
	Al	1850	.8	2.5	5.8	47.2	30.9	12.9
		1860	.1	.5	3.7	23.0	32.2	30.4
	NS	1850	.7	3.1	6.2	46.3	29.4	14.3
		1860	.5	3.2	7.2	62.6	14.1	12.4

TABLE 44 (Continued)

Item	Region	Year	Less Than 20	20-49	50-99	100-499	500-999	1000 & Larger
			Farm Size (Improved Acreage)					
Slave Females	OS	1850	1.6%	2.3%	7.6%	50.7%	24.6%	13.1%
		1860	.2	1.9	3.3	41.5	26.5	26.7
	Al	1850	.7	3.7	6.2	46.4	30.8	12.2
		1860	.1	.9	3.6	33.6	32.4	29.2
	NS	1850	.8	2.6	7.0	44.7	28.7	16.2
		1860	1.4	3.6	8.0	60.3	15.2	11.5
Total Population	OS	1850	2.6	7.8	14.7	48.0	18.1	8.8
		1860	1.2	5.4	7.5	42.3	20.8	22.9
	Al	1850	2.8	8.9	9.8	44.3	24.4	9.6
		1860	2.0	3.6	6.5	33.6	29.2	25.1
	NS	1850	3.4	16.1	13.6	38.5	18.4	9.9
		1860	5.6	12.1	15.2	50.2	9.7	7.3

Source: Calculated from 1850 and 1860 manuscript census samples.

totals in Table 45. In the Old South region, the percentage of crops produced and wealth owned by farmers with fifty or more slaves increased between 1850 and 1860. However, the share of crops produced and wealth owned by farmers with one hundred slaves or more declined slightly. The share of crop production and wealth accounted for by farmers with one to four slaves declined markedly in the New South between 1850 and 1860. In the Alluvial region, the percentage of crops produced and wealth owned by slaveholders with one hundred slaves or more increased during the decade. In the New South, the most noticeable change during the period is the decline in the share of wealth owned and crops produced by farmers with fifty slaves or more. Once again, we find no evidence that the larger planters were displacing the smaller farmers in the New South. However, it does appear that the larger planters in the Alluvial and Old South regions increased their share of crop production and wealth holdings during the 1850's.

SUMMARY AND CONCLUSIONS

The above analysis of the 1850 and 1860 data fails to support the hypothesis that, in the newer regions, the smaller farmers were being bought out by the larger planters.

In the first comparison of the 1850 and 1860 data, we observed that the only statistically significant census-to-census difference in the size distribution of farms occurred in the Old South region. In this region there was a significant increase in the percentage of farms in the larger acreage classes. There appeared to be an increase in the percentage of larger farms in the Alluvial region between 1850 and 1860 but the 1850-1860 differences in the farm-size distributions were not significant.

TABLE 45

CROP OUTPUT, WEALTH, ACREAGE, AND POPULATION BY ACREAGE CLASSES AS A PERCENT OF REGIONAL TOTALS--1850 VS. 1860

			Size of Slaveholding						
Item	Region	Year	0	1-4	5-9	10-19	20-49	50-99	100 & Larger
No. of Farms	OS	1850	27.8%	24.9%	19.1%	14.6%	10.77%	1.9%	1.0%
		1860	29.2	13.3	16.8	17.5	15.0	6.9	1.3
	Al	1850	29.3	14.8	12.3	17.8	17.0	6.8	1.9
		1860	26.9	12.0	13.0	13.5	21.6	7.7	5.3
	NS	1850	45.7	17.9	9.4	13.5	7.6	5.4	.4
		1860	49.0	13.2	12.5	12.8	9.9	2.0	.7
Improved Acreage	OS	1850	8.2	11.3	12.8	19.4	29.6	8.9	9.7
		1860	6.7	4.9	11.0	16.6	29.1	23.6	8.1
	Al	1850	7.9	5.0	5.4	15.4	28.0	27.4	11.0
		1860	3.5	3.1	5.3	9.7	30.8	20.6	26.8
	NS	1850	13.5	9.3	6.7	18.4	16.9	32.1	3.1
		1860	15.4	7.7	10.2	19.3	29.2	11.7	6.5
Unimproved Acreage	OS	1850	9.6	12.1	13.4	15.6	29.4	3.9	16.0
		1860	10.1	5.5	10.3	13.5	29.6	26.4	4.6
	Al	1850	7.6	5.8	6.5	15.2	24.2	22.0	18.7
		1860	7.4	4.6	12.2	12.6	23.9	14.2	26.1
	NS	1850	22.2	16.4	7.0	23.9	10.7	16.6	3.1
		1860	23.1	11.1	11.6	25.4	22.7	4.0	2.1
Val. of Farm	OS	1850	6.0	10.8	11.3	20.5	27.7	10.4	13.3
		1860	6.9	5.3	11.3	17.1	29.2	24.7	5.5
	Al	1850	3.9	3.1	3.5	10.3	25.0	26.6	27.7
		1860	2.1	1.9	3.2	6.8	21.4	19.9	44.6
	NS	1850	9.5	8.0	5.7	16.2	15.3	37.6	7.7
		1860	10.5	6.3	7.8	23.0	29.0	12.1	11.3

TABLE 45 (Continued)

Item	Region	Year	0	1-4	5-9	10-19	20-49	50-99	100 & Larger
			Size of Slaveholding						
Swine	OS	1850	12.1%	15.0%	16.3%	20.1%	22.5%	6.3%	7.6%
		1860	11.9	7.3	11.5	18.6	23.3	22.3	5.1
	Al	1850	16.5	8.3	10.3	18.0	27.4	14.9	4.7
		1860	13.6	7.7	8.0	13.8	26.8	15.2	14.9
	NS	1850	24.9	13.1	7.5	18.8	13.8	20.8	1.0
		1860	22.1	9.5	10.7	24.4	23.6	8.0	1.7
Corn	OS	1850	9.3	11.9	15.5	20.0	26.7	8.0	8.6
		1860	7.7	4.9	11.2	15.4	31.2	21.4	8.4
	Al	1850	11.2	5.6	8.4	15.0	26.4	23.0	10.4
		1860	5.4	4.2	5.7	9.3	34.4	18.0	23.0
	NS	1850	14.8	10.7	8.4	17.9	16.3	29.1	2.9
		1860	16.1	7.6	9.4	23.5	25.6	10.0	7.7
Tobacco	OS	1850	---	---	---	28.6	71.4	---	---
		1860	71.4	---	28.6	---	---	---	---
	Al	1850	50.9	13.6	15.8	19.8	---	---	---
		1860	---	---	---	---	---	---	---
	NS	1850	85.7	14.3	---	---	---	---	---
		1860	96.0	4.0	---	---	---	---	---
Cotton	OS	1850	4.7	7.2	12.0	19.0	28.5	13.2	15.4
		1860	4.6	4.2	8.7	15.2	27.1	28.0	12.0
	Al	1850	3.0	2.4	3.1	13.2	28.3	33.6	16.5
		1860	2.1	1.9	3.7	7.6	26.5	22.5	35.8
	NS	1850	5.9	7.4	6.5	15.7	17.7	37.9	8.8
		1860	7.8	4.3	5.4	22.1	31.2	19.8	9.4

TABLE 45 (Continued)

Item	Region	Year	0	1-4	5-9	10-19	20-49	50-99	100 & Larger
			Size of Slaveholding						
Sweet Potatoes	OS	1850	12.1%	12.1%	16.7%	24.5%	20.8%	5.5%	8.3%
		1860	9.7	5.8	11.8	14.2	24.6	24.2	9.7
	Al	1850	9.5	8.1	8.5	22.3	27.3	19.7	4.7
		1860	12.7	3.4	6.6	7.5	35.6	18.3	15.9
	NS	1850	15.7	9.6	9.2	27.4	13.6	21.7	2.7
		1860	21.1	9.5	7.8	21.8	27.5	10.6	1.6
Free Males	OS	1850	26.1	25.2	19.2	14.4	12.2	2.2	.6
		1860	29.9	13.3	17.7	16.4	14.6	7.1	1.0
	Al	1850	29.4	15.2	10.6	19.9	16.7	6.3	1.8
		1860	34.1	8.7	13.0	11.7	18.8	9.8	4.0
	NS	1850	48.8	19.2	8.4	11.4	6.4	5.0	.8
		1860	50.2	13.0	12.8	11.9	10.0	1.4	.8
Free Females	OS	1850	29.8	25.0	19.3	12.6	10.9	1.8	.6
		1860	30.7	11.9	14.6	18.4	13.6	8.0	2.8
	Al	1850	32.5	15.9	12.5	15.9	16.0	5.8	1.5
		1860	30.3	10.1	12.6	12.2	22.5	9.5	2.9
	NS	1850	49.0	17.3	9.2	11.9	6.5	5.7	.3
		1860	50.5	14.1	12.9	12.1	8.0	1.3	1.1
Slave Males	OS	1850	---	4.2	12.8	21.4	35.2	13.4	13.1
		1860	---	2.0	6.9	15.6	31.1	33.8	10.6
	Al	1850	---	1.8	5.1	14.6	32.3	28.8	17.5
		1860	---	1.0	4.1	8.9	28.4	22.6	35.1
	NS	1850	---	4.3	6.3	21.0	22.5	39.7	6.2
		1860	---	2.8	9.2	22.4	38.4	16.6	10.6

TABLE 45 (Continued)

Item	Region	Year	Size of Slaveholding 0	1-4	5-9	10-19	20-49	50-99	100 & Larger
Slave Females	OS	1850	---	6.0%	14.8%	20.5%	31.7%	13.4%	13.7%
		1860	---	2.2	8.2	16.6	31.8	30.7	10.7
	Al	1850	---	2.2	5.3	15.0	31.6	29.5	16.2
		1860	---	1.3	3.4	8.4	31.0	21.2	34.8
	NS	1850	---	4.1	7.3	20.3	22.3	38.7	7.3
		1860	---	4.4	11.5	21.4	36.1	17.8	9.0
Total Population	OS	1850	10.2	12.4	15.8	18.2	25.4	9.2	8.7
		1860	8.5	5.0	10.0	16.5	26.5	25.2	8.2
	Al	1850	8.1	5.6	6.8	15.7	27.9	23.1	12.9
		1860	5.6	2.6	5.3	9.2	28.1	19.7	29.4
	NS	1850	20.1	10.0	7.6	17.0	15.9	25.3	4.2
		1860	21.1	7.8	11.4	17.7	25.4	10.5	6.0

[a]The 1850 and 1860 samples each contained only one farm having more than 200 slaves.

Source: Calculated from 1850 and 1860 manuscript census samples.

Secondly, we observed that removal of the biases in the samples did not significantly alter our findings. Unfortunately, the effect of sample bias could only be removed when considering the size distributions of farms. However, it seems reasonable to assume that, since the sample biases have little effect on the comparisons of farm-size distributions, their effect on the other distributions is not so great as to significantly alter our conclusions.

Our third observation was based on a comparison of the size distributions of regional slaveholdings (including nonslaveholders) in each of the census years. Again, the only census-to-census differences which were significant (at the .15 level) occurred in the Old South where there was an increase in the percentage of larger slaveholdings between 1850 and 1860. In the Alluvial and New South regions, the 1850 and 1860 size distributions of slaveholdings were not significantly different. However, in the Alluvial region there appeared to be an increase in the percentage of larger slaveholders and a decrease in the percentage of nonslaveholders. In the New South, the percentage of nonslaveholders increased and the percentage of larger slaveholders decreased.

Fourth, we observed that if nonslaveholders were excluded from the size distribution of slaveholdings the only significant census-to-census difference again occurred in the Old South region where the percentage of larger slaveholdings increased markedly. Although not statistically significant, there was an increase in the percentage of Alluvial slaveholdings of twenty slaves or more between the two census years.

Fifth, we observed that, in the Old South region, the percentage of crops produced and wealth owned by farmers with more than five hundred improved acres increased between 1850 and 1860 as did the percentages of the free and slave population living on these farms. In the Alluvial region, the changes were the same as in the Old South but in the New South the findings were just the reverse. In the New South the share of wealth held and crops produced by the larger farmers declined during the 1850's.

Finally, in the Old South region, the share of crops produced and wealth owned by farmers with fifty or more slaves increased between 1850 and 1860 although the share accounted for by farms with one hundred slaves or more declined slightly. In the Alluvial region, the share of crop outputs and wealth accounted for by farms with one hundred slaves or more increased during the decade while, in the New South, the share accounted for by larger slaveholders declined.

Before drawing any conclusions, it should again be pointed out that the results of this analysis are applicable to the cotton growing regions only. Rice, sugar, and tobacco producing regions are not adequately represented in the manuscript census samples. In fact, some readers may insist that the generalizations of this chapter apply only to the seventy-four major cotton producing counties from which the sample is selected.

Now, what conclusions can be drawn from our observations? First, and, for the purposes of this chapter, foremost, the observations clearly <u>fail</u> to support the second part of our hypothesis. The lack of a statistically significant census-to-census difference in the size distribution of farms in the New South region and the lack of a statistically significant census-to-census difference in size distribution of slave-holdings in the New South region between 1850 and 1860 fail to support the hypothesis that the large planters were forcing the small farmers out of the better lands in the New South. In fact, the increase in the percentage of large farms in the Old South region, the increase in the percentage of large slave holdings in the Old South region, and the increase in the importance of large farms and large slaveholdings in Old South crop production and wealth holdings indicate that the small farmer was being forced out of the Old South rather than the New South region. Only in the Alluvial region do we find any support for our hypothesis. In this region, we do find an increase in the percentage of large farms and large slaveholders. Further, we observe that the large farms and large slaveholders became relatively more important producers of crops and holders of wealth during the decade of the 1850's. However, these tendencies are not statistically significant. Thus, it seems clear that the second part of our hypothesis is not applicable to the Southwest taken as a whole.

Our second conclusion might well be that the hypothesis is applicable to the Alluvial region for it does appear possible that the large planters had bought out the small farmers in this region. At least, we have a clue to the origin of our hypothesis since a myopic view of southern agriculture which focused only on the Alluvial region might well result in such an assumption.

Finally, the results of this chapter cast further doubt on the validity of the first part of our hypothesis. Clearly, if the size distributions of farms and the size distributions of slaveholdings in the New South remained constant during a period in which small farmers were being "squeezed out" of the Old South, it seems unreasonable to conclude that large farmers were leading the migration to the New South. It seems much more likely that the small farmers sold their land and slaves to large planters and then moved to the Southwest. The fact that smaller farmers and smaller slaveholders were more predominant in the New South than in the Old South in both 1850 and 1860 lends further weight to this conclusion.

At this point, the reader may become tempted to raise at least two questions. First, he may ask whether it is possible that the larger planters were just beginning to buy out the smaller farmers in 1860. That is, he may question whether the period covered in this study truly enables us to evaluate the second part of our hypothesis. Clearly, this question can not be fully answered. However, the writer is tempted to remark that, due to the outbreak of the Civil War, any buying out of small farmers by large planters had to occur during the decade of the fifties or before. That is, any generalizations in support of the hypothesis clearly must depend on pre-1860 data. Therefore, it seems appropriate to test the hypothesis using data from the same period. Further, by selecting data from near the end of the antebellum era, we have given the planters every opportunity to make their purchases.

A second question the reader may wish to raise relates to land qualities in the Alluvial and New South regions. Because the second part of our hypothesis suggests that the small farmers were pushed out of the "better land", the reader may be tempted to maintain that it is reasonable to observe that the small farmers were pushed out of the Alluvial region but not out of the New South region. In fact, the reader may want to assert that the small farmers who were pushed out of the Alluvial regions may have fled to the New South region. However, it should be recalled that the New South region included the rich lands of the Alabama "black belt" and each of the New South counties is cited by Gray as having been a major cotton producing county in 1859. That is, the New South counties were not frontier counties and, further, they were counties which contained better quality land. Thus, the hypothesis should apply to the New South as well as to the Alluvial.

The data in the published census volumes and in our manuscript census samples enable us to shed some light on the proposition that the smaller farmers were being pushed off the "better land". Farm values per improved acre and per acre (improved and unimproved) and the ratio of improved to unimproved acreage are presented in Table 46. Farms with less than twenty improved acres had a value per <u>improved</u> acre which was greater than the regional average in the Old South in 1850 and 1860, in the Alluvial region in 1850 and in the New South in 1850 and 1860. However, the value per acre (improved <u>and</u> unimproved) for these same farms exceeded the regional average only in the Old South in 1850. Obviously, this difference results from the fact that farms with less than twenty improved acres had a ratio of improved to unimproved acres which was much lower than that on larger farms. Because of this, one would expect farms with less than twenty improved acres to have values per improved acre which were higher than the regional averages and values per acre (improved and unimproved) which were below the regional averages if the land qualities on large and small farms were comparable.

A glance at Table 46 indicates that the results do not always conform to our expectations. First, the quality of land on Old South farms with less than twenty acres appears to have been better than average in both 1850 and 1860. (For 1850, this is obvious; for 1860, this conclusion is not unequivocal, yet appears quite secure.) In the Old South the apparent superiority of land on small farms seems to have been reduced between 1850 and 1860. Second, in the Alluvial region,

TABLE 46

FARM VALUES PER ACRE--OLD SOUTH, ALLUVIAL AND NEW SOUTH--1850 AND 1860

	Old South		Alluvial		New South	
	1850	1860	1850	1860	1850	1860
Value of Farm/Improved Acre:						
Region (Published Census)	$17.74	$22.60	$24.34	$63.40	$15.33	$33.99
Region (Sample)	18.47	20.35	29.03	74.54	14.17	30.12
Farms 19 acres	30.41	48.87	34.30	33.92	19.01	45.35
Value of Farm/Acreage (Improved and Unimproved):						
Region (Published Census)	5.71	9.07	8.72	21.89	4.84	10.85
Region (Sample)	6.50	8.39	9.49	25.51	4.67	9.03
Farms 19 acres	7.11	8.29	5.68	5.37	2.43	5.10
Ratio: Improved Acreage/Unimproved Acreage:						
Region (Published Census)	.474	.671	.559	.527	.462	.469
Region (Sample)	.543	.702	.486	.520	.492	.428
Farms 19 acres	.305	.204	.198	.188	.147	.127

Sources: 1850--Calculated from the 1850 manuscript census sample and U.S. Bureau of the Census, Seventh Census of the United States: 1850, Compendium.

1860--Calculated from the 1860 manuscript census sample and U.S. Bureau of the Census, Eighth Census of the United States: 1860, Agriculture.

these smaller farmers apparently held land of approximately average quality in 1850 but, by 1860, they held land which was decidedly inferior. (Despite having a ratio of improved to unimproved acreage only one-third the regional average, smaller farms were valued at one-half the average value per improved acre.) Finally, in the New South smaller farms apparently consisted of land of approximately average quality.

Thus, we can conclude that, in the Alluvial regions of the Southwest, the small farmer was being pushed off the better quality land. In the Old South, the small farmer appears to have held land of better than average quality in both 1850 and 1860. Thus, he could have been pushed off the "better land", perhaps, but clearly all the better land in the Old South was not purchased from the smaller farmers prior to 1860. However, it should be noted that in 1850 land prices in the Old South exceeded those in the New South and the ratios of improved to unimproved acreage were approximately the same. Consequently, any conclusion that small farmers were being pushed out of the Old South may indicate that they were being pushed out of the "better land". By 1860, the value of New South land was greater than Old South land and the displacement of smaller farmers may have ceased.

In summary then, the 1850-1860 comparisons utilized in this chapter fail to support either part of our hypothesis unequivocally. However, we do find evidence that between 1850 and 1860 the smaller farmer had been displaced from the better land in the Alluvial region and from the Old South.

CHAPTER VIII

THE ROLE OF THE SMALL FARMER

In the previous chapters we have endeavored to ascertain the role of the small farmer in westward migration in the antebellum South. We have concluded that large planters clearly did <u>not</u> lead the migration and, in fact, their participation in the migration was not in proportion to their numerical representation in the Old South. Thus, the migration appears to have been predominantly a movement of small farmers and non-slaveholders. Only when the alluvial regions were considered separately, did the larger planters appear to have participated in the migration in proportion to their numerical representation in the Old South.

Our second conclusion was that the smaller farmer does not appear to have been pushed out of the better lands in the Southwest. In fact, more "pushing-out" appears to have occurred in the older southeastern regions. However, we did find that, in the alluvial regions, small farmers may, indeed, have been pushed out by larger planters who were able to buy the better land from them.

In short, it appeared that the small farmer was generally able to hold his own in the cotton-growing regions of the Southwest prior to 1860. In fact, the small farmer may have considered the new cotton-lands in the Southwest to be a place where he could start his quest for prosperity afresh. If this were indeed the case, it would appear that the small farmers who moved into the new cotton lands in the Southwest were more similar to their northern counterparts than they have generally been pictured.

In this chapter, we shall attempt to ascertain the role of the small farmer as a producer of crops, etc. in the primary cotton-producing regions in 1850 and 1860. That is, we shall consider the smaller farmers' participation in various regions rather than their movement between regions.

Before beginning this task, we might remind the reader that the following observations are based on samples of farms selected from the 1850 and 1860 manuscript censuses of agriculture. Further, the samples were selected from the major cotton producing regions only. Thus, any conclusions must be qualified by noting that they are based on sample rather than universe data. Also, only the 1850 and 1860 census years are considered. Finally, the results obtain to cotton producing regions only. That is, if we find that small farmers were primarily herders of cattle and drovers of hogs, we are not saying that small farmers tended to locate on western grazing lands. Rather, we are saying that, in the cotton growing regions, smaller farmers concentrated on the production of cattle and hogs.

The discussion in the following section will be divided into three parts: an analysis based on the 1850 sample; an analysis using the 5229 farm 1860 sample; and an analysis utilizing the three regional sub-samples of the 1860 sample.

ANALYSIS OF THE SAMPLE DATA

The 1850 Sample

As was mentioned in earlier chapters, the 1850 sample was selected from Gray's seventy-four major cotton producing counties and includes only a limited amount of data. Thus, the conclusions drawn from the sample will be somewhat limited.

The crop outputs, acreage, and population on farms of less than fifty improved acres expressed as a percentage of region totals are presented in Table 47. In the New South, these smaller farmers owned 4.7 percent of the improved acreage, 8.9 percent of the unimproved acreage and housed 10.4 percent of the population. They produced only 3.8 percent of the region's cotton, 10.6 percent of its swine, 7.5 percent of its corn and 11.6 percent of its sweet potatoes. Thus, these smaller farmers appear to have produced a proportion of hogs and sweet potatoes greater than their proportion of acreage or population and corn and cotton in lesser proportions. We can conclude that either the people and hogs on smaller farms ate less corn per capita than did people and hogs on larger farms, or they purchased corn from their larger neighbors, or their larger neighbors sold corn to nonfarmers. An additional possibility is that small farmers fed their hogs less corn than did larger farmers perhaps allowing the hogs to roam the unimproved acreage in search of mast. In any case, we can conclude that hogs and sweet potatoes were relatively more important food crops for small farmers than was corn. Additionally, the small farmer produced a smaller percentage of the Old South's cotton than his share of the improved acreage would have suggested. This may be explained by the fact that, because 10.4 percent of the population were living on 4.7 percent of the improved land, a greater share of the improved land had to be used for the production of food crops.

TABLE 47

CROP OUTPUTS ON FARMS OF LESS THAN FIFTY ACRES
OLD SOUTH VS ALLUVIAL VS NEW SOUTH--1850
(Percentage of Regional Totals)

	Old South	Alluvial	New South
Number of farms	26.8	35.3	43.0
Improved acreage	4.7	5.1	9.2
Unimproved acreage	8.9	7.9	23.9
Value of farm	5.6	4.1	7.3
Cotton	3.8	2.7	4.9
Tobacco	0.0	33.8	85.7
Cane sugar	No Production on Sampled Farms		
Molasses	No Production on Sampled Farms		
Swine	10.6	17.4	23.3
Corn	7.5	10.6	13.3
Sweet Potatoes	11.6	10.9	15.9
Free population	22.7	34.3	42.3
Slave population	3.3	3.8	3.7
Total population	10.4	11.7	19.5

Source: Calculated from 1850 manuscript census sample.

In the Alluvial region, the results are basically the same except that in this region even more emphasis appears to have been placed on the production of hogs and less emphasis on the production of cotton. Again, the percentage of corn production is below the percentage of either persons or hogs. The alluvial farmer with less than fifty acres held only 5.1 percent of the improved land but, produced over one-third of the regions' tobacco. Similar findings obtain in the New South. In this region, however, the share of unimproved acreage held by smaller farmers and the percentage of total population on these farms is striking. Again, tobacco appears to have been a relatively important cash crop.

The percentage figures are not the whole story as a glance at the absolute amounts of cotton and tobacco produced in the regions will indicate. Tobacco and cotton production on farms with less than fifty acres by region were: Old South--no tobacco, 195 bales cotton; Alluvial --685# tobacco, 341 bales cotton; New South--300# tobacco, 169 bales cotton. William Brown, an assistant marshall who recorded the 1850 census data in Newton County, Georgia, was so kind as to leave us a schedule of prices for the produce of his region. Mr. Brown valued tobacco at seven cents per pound and 400-pound bales of cotton at $36.40 per bale.[1] Using Mr. Brown's prices, the cash contributions of cotton and tobacco to the farms with less than fifty acres in each of the regions were: Old South--cotton, $7,098.00, tobacco, $0.00; Alluvial--cotton, $12,412.40, tobacco, $47.95; New South--cotton, $6,151.60, tobacco, $21.00. Clearly, cotton, not tobacco, was the cash crop of smaller farmers in the primary cotton growing regions during the 1849 crop year.

Crop outputs, acreage, and population on slaveless farms are expressed as a percentage of regional totals in Table 48. The same data, for farms with one to nine slaves are presented in Table 49. For slaveless farms, the results are similar to those for farms with less than fifty improved acres. The percentages of foodstuffs produced exceeded the percentages of improved acreage owned and the percentage of cotton grown was less than the percentage of improved acreage in each region. In the Alluvial

---

[1]A check with Arthur Harrison Cole, Wholesale Commodity Prices in the U.S., 1700-1861 (Cambridge: Harvard University Press, 1938)indicates that Mr. Brown's prices were quite reasonable. For 1849 and 1850, Cole's cotton prices range from $.06 to $.15 per pound ($24.00 to $60.00 per 400# bale) and his tobacco prices from $.04 to $.105 per pound.

TABLE 48

CROP OUTPUTS ON SLAVELESS FARMS
OLD SOUTH VS ALLUVIAL VS NEW SOUTH--1850
(Percentage of Regional Totals)

	Old South	Alluvial	New South
Number of farms	27.8	29.3	45.7
Improved acreage	8.2	7.9	13.5
Unimproved acreage	9.6	7.6	22.2
Value of farm	6.0	3.9	9.5
Cotton	4.7	3.0	5.9
Tobacco	0.0	50.9	85.7
Cane sugar	No Production on Sampled Farms		
Molasses	No Production on Sampled Farms		
Swine	12.1	16.5	24.9
Corn	9.3	11.2	14.8
Sweet potatoes	12.1	9.5	15.7
Free population	27.9	30.8	48.9
Slave population	--	--	--
Total population	10.2	8.1	20.1

Source: Calculated from 1850 manuscript census sample.

TABLE 49

CROP OUTPUTS ON FARMS WITH ONE TO NINE SLAVES
OLD SOUTH VS ALLUVIAL VS NEW SOUTH--1850
(Percentages of Regional Totals)

	Old South	Alluvial	New South
Number of farms	44.0	27.1	27.3
Improved acreage	24.1	10.4	16.0
Unimproved acreage	25.5	12.3	23.4
Value of farm	22.1	6.6	13.7
Cotton	19.2	5.5	13.9
Tobacco	0.0	29.4	14.3
Cane sugar	No Production on Sampled Farms		
Molasses	No Production on Sampled Farms		
Swine	31.3	18.6	20.6
Corn	27.4	14.0	19.1
Sweet potatoes	28.8	16.6	18.8
Free population	44.3	27.0	27.1
Slave population	18.8	7.2	11.0
Total population	28.2	12.4	17.6

Source: Calculated from 1850 manuscript census sample.

region the percentage of corn produced exceeded the percentage of population living on these farms. In the other regions corn production per capita was lower than the average for all farms in those regions. Hog production per capita was above the regional average in each of the regions. In the Alluvial region hog production per capita was twice the regional average thus accounting for the unusually high corn production in this region.

On farms with one to nine slaves similar results obtain. However, one interesting difference may be noted in Table 49. In the Old South and New South regions, farmers with one to nine slaves produced cotton almost in proportion to their improved acreage holdings and their production of foodstuffs was only slightly above their proportion of population. In the Alluvial region, however, farmers with one to nine slaves produced only one-half the average amount of cotton per improved acre and raised hogs, corn and sweet potatoes in quantities greater than proportional to the population living on the farms.

Before leaving the 1850 data, one final observation should be made. In the New South, farmers with less than fifty improved acres, farmers without slaves, and farmers with one to nine slaves held approximately twice the percentage of unimproved land as improved land. Especially since the number of unimproved acres in the New South was twice the number of improved acres, these large unimproved acreage holdings are striking.

The 5229-farm 1860 Sample

The reader will recall that in earlier chapters we divided the 5229-farm 1860 sample into two parts: Old South states east of the

Alabama-Georgia border and New South states to the west of that border. This division is maintained in the present section.

Crop outputs, acreage, animal inventories and population on farms with less than fifty acres, on slaveless farms and on farms with one to nine slaves are expressed as percentages of regional totals in Table 50. Farms of less than fifty acres in 1860 comprised 28.5 percent of the Old South farms and 45.4 percent of the New South farms but included only 4.9 percent of Old South and 9.9 percent of New South improved acreage. Their share of the unimproved acreage was twice their share of the improved land. In the Old South, these smaller farmers appear to have owned land which was comparable in value to the land held by other farmers but in the New South the smaller farmers' holdings had a value per acre below the regional average. (Old South farmers with less than fifty improved acres held 9.9 percent of the improved acreage and 21.9 percent of the unimproved acreage yet their farms were valued at only 8.2 percent of the regional total.)

On these smaller farms, free persons, horses, and oxen appear to have contributed a greater percentage of the labor than on larger farms while asses and mules contributed a lesser proportion. In both regions, these smaller farmers raised milk cows, oxen, other cattle, and swine in proportions greater than their share of improved or unimproved acreage and, in the New South, raised a greater proportion of sheep also. The lower proportion of the value of livestock held on these smaller farms, especially in the New South region, indicates that the livestock owned by smaller farmers must have been of inferior quality or at least was given a lower value by the census enumerator. (In the New South, the only thing which could make this last statement false would be for the

TABLE 50

CROP OUTPUTS ON FARMS OF LESS THAN FIFTY ACRES, ON SLAVELESS FARMS AND ON FARMS WITH ONE TO NINE SLAVES--OLD SOUTH VS NEW SOUTH--1860
(Percentage of Regional Totals)

	Under Fifty Acres		Slaveless Farms		One to Nine Slaves	
	Old	New	Old	New	Old	New
Number of farms	28.5	45.4	41.3	55.5	32.2	23.6
Personal estate	4.1	6.6	3.1	5.5	16.8	15.5
Improved acreage	4.9	9.9	13.1	19.6	22.3	17.9
Unimproved acreage	9.9	21.9	16.9	25.5	25.0	25.0
Value of farm	5.8	8.2	12.0	11.1	22.3	14.1
Horses	15.2	28.0	26.0	35.8	31.1	23.2
Asses & mules	4.7	12.8	8.5	16.4	18.6	13.4
Milk cows	15.6	30.2	23.6	36.5	27.6	22.6
Working oxen	10.8	25.2	19.4	33.8	29.1	21.6
Other cattle	12.9	28.2	19.6	31.3	28.3	21.2
Sheep	8.4	29.9	20.0	37.4	30.2	32.3
Swine	13.3	24.4	21.5	32.0	27.3	23.2
Value of livestock	9.6	18.1	16.3	23.7	25.0	20.4
Value of animals slaughtered	11.4	21.1	19.7	29.2	25.1	20.8
Cotton	4.3	6.3	8.4	8.9	16.6	11.6
Tobacco	8.0	34.6	13.3	39.7	45.5	42.1
Cane sugar	0.0	0.0	1.7	0.0	1.6	0.0
Molasses	10.9	1.3	20.9	2.4	18.1	.9
Corn	8.7	14.9	16.3	22.5	24.8	19.0
Peas & beans	12.5	17.7	19.3	22.8	22.7	18.1
Irish potatoes	13.2	25.8	19.5	30.0	22.7	19.7
Sweet potatoes	14.1	20.3	22.8	26.3	25.7	19.4
Home manufacturing	18.8	34.7	31.7	50.1	29.0	20.4
Free population	26.5	43.6	43.5	58.0	31.0	22.9
Slave population	2.7	3.3	---	---	15.1	13.2
Total population	12.1	21.7	17.1	26.5	21.4	17.7

Source: Calculated from 1860 manuscript census sample.

low percentage of asses and mules held, 12.8 percent, to have been of significant importance to lower the average value of livestock to the 18.1 percent level. Considering that the New South sample contained only 8,555 asses and mules compared with 13,817 horses, 124,105 swine, etc., this would have been impossible.)

As was the case in 1850, farms with less than fifty improved acres produced a smaller percentage of cotton and a greater percentage of tobacco than their share of improved acreage would have indicated. Again, however, cotton proves to have been the cash crop. Even using Cole's minimum cotton price and maximum tobacco price for 1859 and 1860, the cash contribution of cotton far exceeded the cash contribution of tobacco. (Approximate figures: Old South--cotton, $32,700, tobacco, $498; New South--cotton, $145,200, tobacco, $5,465.)[2] The percentage of corn production on smaller farms in each of the regions exceeded the percentage of improved acreage but was less than the percentage of population or livestock on the farms. Again, the smaller farmers either consumed proportionately less corn or purchased it from the larger farmers or these larger farmers sold corn to nonfarmers. Farmers with less than fifty acres produced a larger share of each region's peas and beans, Irish potatoes and sweet potatoes than their share of improved acreage and their production of these foodstuffs was approximately proportional to their share in the total population. As in 1850, the smaller farmer apparently found it necessary to devote a greater proportion of his improved acreage to the production of food than did larger farmers.

---

[2] Ibid.

The picture on slaveless farms was little different from that on farms of less than fifty improved acres with two notable exceptions. First, on slaveless farms the percentage of unimproved acreage was only about one-third larger than the percentage of improved acreage held. Second, in both the Old and New South, slaveless farmers owned land which was lower in value than the regional average. As on farms with less than fifty acres, slaveless farmers farmed with horses or oxen and tended to shun asses and mules. The quality of their livestock holdings was either inferior to the regional average or was undervalued by the enumerators. Finally, they appear to have put relatively more emphasis on the production of foodstuffs than on the production of cotton.

A striking feature of both slaveless farms and farms with under fifty improved acres was the large percentage of home manufactures which were produced on these farms. It appears that the value of home manufacturing was approximately proportional to the free population living on the farms. In fact, this correspondence between the percentage of home manufactures and the percentage of free population living on the farm is also evident on farms with one to nine slaves.

In some other respects, farms with one to nine slaves differed from slaveless farms and farms with less than fifty acres. In the Old South, farms with one to nine slaves held only a slightly larger share of the unimproved acres than the improved acres. Also, the percentage of farm value represented by these farms was approximately the same as the percentages of improved and unimproved acreage owned. Thus, in the Old South, the farmers with one to nine slaves appear to have held land of average quality. Farms with one to nine slaves tended to produce a larger proportion of livestock (excepting asses and mules), corn and other food crops and a smaller proportion of cotton than their share of improved acreage would suggest. However, the percentage of corn produced on these farms exceeded the percentage of total population and is somewhat less than the percentage of livestock owned. Perhaps the greater percentage of slaves on these farms, as compared to slaveless farms and farms with less than fifty improved acres, was the cause of the increase in corn per capita.

The 1860 Regional Subsamples

The regional subsamples of the 1860 manuscript census sample were selected from Gray's three major cotton producing regions--Old South, Alluvial, and New South--defined earlier.

The crop outputs, acreage, animal inventories, and population on farms with less than fifty improved acres in the three regions are expressed as percentages of regional totals in Table 51. The conclusions which may be drawn from Table 51 are similar to those drawn from the Old South-New South breakdown in Table 50. The percentage of unimproved acreage on these farms was twice the percentage of improved acreage. Free persons, horses, and oxen were a relatively greater part of the labor supply on these farms than on larger farms; asses and mules, a relatively lesser part. The large percentages of animal inventories (other than asses and mules and sheep) are striking. However, it again appears that either the animals on these farms were of inferior quality or they were undervalued. The percentage of cotton produced was lower than the percentage of improved acreage held but cotton, again, was the major cash crop in all three regions. The percentage of corn produced exceeded the percentage of improved acreage but was considerably below

TABLE 51

CROP OUTPUTS ON FARMS OF LESS THAN FIFTY ACRES
OLD SOUTH VS ALLUVIAL VS NEW SOUTH--1860
(Percentage of Regional Totals)

	Old South	Alluvial	New South
Number of farms	19.5	26.4	40.1
Personal estate	2.6	1.5	8.6
Improved acreage	1.9	2.4	7.7
Unimproved acreage	4.7	6.4	14.4
Value of farm	3.3	1.0	7.7
Horses	8.6	10.5	24.2
Asses & mules	2.0	2.1	7.4
Milk cows	7.3	10.0	27.5
Working oxen	6.9	6.6	15.1
Other cattle	5.4	11.2	32.3
Sheep	3.8	2.3	6.8
Swine	6.9	11.2	21.1
Value of livestock	4.8	5.0	14.6
Value of animals slaughtered	5.8	9.1	19.6
Cotton	1.8	1.0	5.9
Tobacco	71.4	---**	40.0
Cane sugar	0.0	0.0	---**
Molasses	0.9	0.0	11.7
Corn	3.9	3.9	11.2
Peas & beans	7.1	7.1	18.5
Irish potatoes	5.2	8.0	15.7
Sweet potatoes	6.2	8.0	15.7
Home manufacturing	13.6	11.4	28.0
Free population	18.1	27.8	36.4
Slave population	1.9	0.8	4.4
Total population	6.6	5.6	17.7

**No production in sample.

Source: Calculated from 1860 manuscript census sample.

the percentages of total population or livestock living on the farm. A corn deficit again appears possible but the production of other foodstuffs was generally in proportion to the number of people living on these farms.

One conclusion not readily apparent from Table 50 should be noted. As indicated by Table 50, the value per acre of these smaller farms in the Old South region was approximately the same as that of larger farms and the acreage values in the newer regions were lower for small farms than larger ones. This conclusion obtains for both the Alluvial and New South regions. However, in the Alluvial region, the per acre value of smaller farms was <u>much</u> less than that of larger farms (no more than forty percent as great). This supports our earlier observation that, in the Alluvial region, small farmers may have been squeezed off the best lands.

Table 52 presents crop outputs, acreage, etc. on slaveless farms in each of the regions. Again, the conclusions are the same as those obtained from Table 50 with two exceptions. First, the percentage of unimproved acreage on these farms was one and one-half to two times as great as the percentage of unimproved acreage. In the Alluvial region, the percentage of unimproved acreage was over twice as great. Second, the value per acre of slaveless farms in the Old South was approximately the same as that of farms with slaves. In the Alluvial and New South regions, the values were much lower for slaveless farms--no more than sixty to seventy percent of the value for farms with slaves.

Crop outputs, etc. for farms with one to nine slaves are expressed as a percentage of regional totals in Table 53. Because the conclusions which may be drawn from Table 53 are similar to those mentioned in the discussion of Table 50, only one further comment needs to be made:

TABLE 52

CROP OUTPUTS ON SLAVELESS FARMS
OLD SOUTH VS ALLUVIAL VS NEW SOUTH--1860
(Percentage of Regional Totals)

	Old South	Alluvial	New South
Number of farms	29.2	26.9	49.0
Personal estate	1.4	1.2	4.3
Improved acreage	6.7	3.5	15.4
Unimproved acreage	10.1	7.4	23.1
Value of farm	6.9	2.1	10.5
Horses	15.7	12.0	33.1
Asses & mules	3.7	2.0	8.6
Milk cows	13.8	12.2	33.4
Working oxen	10.2	7.7	25.5
Other cattle	11.1	12.1	39.2
Sheep	8.2	3.9	15.9
Swine	11.9	13.6	22.1
Value of livestock	8.8	6.1	18.9
Value of animals slaughtered	12.7	11.3	24.8
Cotton	4.6	2.1	7.8
Tobacco	71.4	---**	96.0
Cane sugar	0.0	0.0	---**
Molasses	0.9	0.0	3.7
Corn	7.7	5.4	16.1
Peas & beans	10.8	7.8	22.2
Irish potatoes	10.1	12.2	20.3
Sweet potatoes	9.7	12.7	21.1
Home manufacturing	20.8	15.9	41.6
Free population	30.6	32.2	50.3
Slave population	---	---	---
Total population	8.5	5.6	21.1

**No production in sample.

Source: Calculated from 1860 manuscript census sample.

TABLE 53

CROP OUTPUTS ON FARMS WITH ONE TO NINE SLAVES
OLD SOUTH VS ALLUVIAL VS NEW SOUTH--1860
(Percentage of Regional Totals)

	Old South	Alluvial	New South
Number of farms	30.1	25.0	25.7
Personal estate	10.3	8.0	17.2
Improved acreage	15.9	8.4	17.9
Unimproved acreage	15.8	16.8	22.7
Value of farm	16.6	5.1	14.1
Horses	25.9	15.7	29.6
Asses & mules	12.4	4.7	13.1
Milk cows	21.2	14.5	22.2
Working oxen	24.7	10.4	23.2
Other cattle	21.0	14.2	20.0
Sheep	29.1	6.5	19.0
Swine	18.8	15.7	20.2
Value of livestock	18.8	9.2	21.0
Value of animals slaughtered	20.3	13.0	18.4
Cotton	12.9	5.6	9.7
Tobacco	28.6	---**	4.0
Cane sugar	0.0	0.0	---**
Molasses	1.6	0.0	96.3
Corn	16.1	9.9	17.0
Peas & beans	16.6	13.6	25.5
Irish potatoes	19.3	10.4	20.1
Sweet potatoes	17.6	10.0	17.3
Home manufacturing	20.7	16.7	25.5
Free population	28.8	22.1	26.4
Slave population	9.6	4.9	13.9
Total population	15.0	7.9	19.2

**No production in sample.

Source: Calculated from 1860 manuscript census sample.

molasses was not a significant cash crop in the New South region (total production in sample--299 gallons).

SUMMARY

Between 1850 and 1860 the percentage of farms with less than fifty improved acres declined in each of the three major cotton producing regions. This decline was especially large in the Alluvial and Old South regions. In each of these regions, the smaller farmer's share of the improved acreage declined much more than his share of the unimproved acreage, a fact which serves to underscore some of the basic characteristics of the smaller farmer's operation. He appears to have been a holder of relatively more unimproved land than improved land. He raised relatively more corn and hogs than cotton. Yet, he either had a corn deficit, bought corn from his larger neighbors, or sold less corn to non-farmers than they did. In any event, the corn shortage probably did little damage for he generally raised cattle, hogs, peas and beans, Irish potatoes and sweet potatoes in a greater proportion than his share of the population and, with a greater share of unimproved land, his hogs could forage for mast to help offset any corn shortage. Since by far the majority of the people on his farm were free rather than slave, it is possible that his corn shortage was fiction, i.e., he didn't need nor want any more corn in his diet. His cash crop was cotton, not tobacco or sugar and his share of each region's home manufacturing exceed his share of the total population (including slaves). His land (except in the Old South) and his livestock were probably inferior to those of his larger neighbors but his labor force consisted proportionately more of free people and horses and proportionately less of slaves and mules.

Between 1850 and 1860, the percentage of slaveless farmers increased in the Old and New South regions but declined slightly in the Alluvial. In the Alluvial the slaveless farmers also held a much greater percentage of unimproved land than improved land. In the other two regions their excess of unimproved land was not so great. The slaveless farmer raised relatively more corn and hogs than cotton. He, too, may have had a corn deficit which was offset by his production of cattle, hogs, peas and beans, Irish potatoes and sweet potatoes. In the Alluvial region his hog production per capita was twice the regional average indicating a possible hog surplus. His cash crop was cotton and his value of home manufacturing relatively high. In the Old South, his land was comparable to other slaveholding farmers but his livestock was probably inferior. In the Alluvial and New South, both were inferior. Of course, his human labor force was free or rented; his animal work force contained relatively more horses and less mules than his slaveholding neighbor's.

The percentage of farms with one to nine slaves declined markedly in the Old South between 1850 and 1860 but remained constant in the other two regions. In the Old South and New South regions in 1850 and in the Old South region in 1860, the farmer with one to nine slaves produced a percentage of cotton about in proportion to his improved acreage holdings and food crops in proportion to the mouths he had to feed. (His acreage holdings were approximately proportional to the population on his farm so he could do both.) In the Alluvial region in 1850 and in the New South and Alluvial region in 1860, the percentage of cotton produced by farmers with one to nine slaves was approximately one-half their percentage of improved acreage. In the Alluvial region, these farmers produced a percentage of corn, hogs and other foodstuffs which exceeded their share of the total population. In 1850, these farmers owned inferior or undervalued land in each of the regions. By 1860, this was the case only in the New South and Alluvial regions, especially in the latter.

Thus, one gets the picture that the smaller farmer generally farmed land of somewhat inferior quality and held a larger share of the unimproved than the improved acreage. On this land he raised relatively more food crops than his larger neighbors did and, consequently, relatively less cotton. His relatively larger proportion of animal holdings, other than sheep and mules, reflects his unimproved acreage holdings and his production of foodstuffs. His animals were of lower quality than his larger neighbors' which may reflect a diet of mast rather than corn. If so, the small farmer may have been a poorer, yet rational, economic man feeding his animals factors relatively cheaper to him. Relative to all southerners, his own table was likely well set with milk, meat, peas, beans, and Irish potatoes for he raised these products in proportions greater than his share of the total population. Compared to other free persons, however, such a conclusion does not necessarily obtain. (Such a comparison requires knowledge of how various foodstuffs were distributed between free persons and slaves.) Cotton was the smaller farmer's cash crop but he may have supplemented his income with sales of foodstuffs and manufactured items.

In short, the small farmer fits in about where he would be expected to. His emphasis on food production was natural in view of his limited land holdings relative to his share of the total population. In the Old South, his land holdings were not inferior; in the New South and Alluvial regions, they clearly were. Perhaps he was improving the land. More likely, he had been squeezed off the best land, especially in the Alluvial region. His life appears to have been clearly superior to that of the average rural southerner. In fact, he appears to fit the niche to which most would assign him--a spot somewhere between planter and slave.

CHAPTER IX

SUMMARY AND CONCLUSIONS

The purpose of this study has been to shed some new light on westward migration in the southern United States. Especially, we have attempted to determine whether this migration is properly characterized as a movement of small farmers _and_ large planters into the newer lands in the Southwest. Also, we have examined the role of the smaller farmer and nonslaveholding farmer in the primary cotton producing regions in both 1850 and 1860.

SUMMARY OF THE STUDY

The study is both a part of and an outgrowth from a larger study of antebellum southern agriculture being undertaken by Professors William N. Parker and Robert E. Gallman and their students at Yale and the University of North Carolina. This larger study is based on a sample of farms selected from the 1860 manuscript census. Initial impetus to the present study was provided by the observation that there appears to have been a greater proportion of nonslaveholders in the Southwest (a newer region) than there was in the Southeast (an older region). Specifically, in the 1860 manuscript census sample selected from Virginia, North Carolina, South Carolina, Georgia and Florida, forty-one percent of the farms were slaveless. On the other hand, fifty-three percent of the sample farms selected from Alabama, Mississippi, Louisiana, and Tennessee and sixty-three percent of those selected from Arkansas and Texas were without slaves. This observation was surprising in view of the fact that several economic historians have characterized southwestward migration as a movement of small farmers _and_ large planters. Further, some writers have indicated that the larger planters purchased the better lands in the Southwest from the smaller farmers forcing them to move on. The larger percentage of nonslaveholders in the more recently settled regions casts some doubt on the validity of the existing characterizations.

Having observed these apparent inconsistencies, the writer has attempted to test the validity of the following two-part hypothesis:

(a) The migration into the new land in the Southwest was one of the large planters leading, accompanying, or _immediately_ following smaller farmers.

(b) In the case that the larger planters immediately followed the smaller farmers, the smaller farmers were pushed out by larger planters who were able to buy the better land from them.

Two statistical samples of farms selected from the 1850 and 1860 manuscript censuses were used to test the hypothesis.

The 1860 sample consists of 5229 farms selected from the 405 counties which produced at least one thousand bales of cotton in the 1859 production year. The 1850 sample contains 897 farms selected from the seventy-four counties which Lewis C. Gray describes as the major cotton producing counties in 1859. Each of these samples was tested to determine whether it is representative of the region from which it was selected using a number of statistical tests. Neither the 1850 nor the 1860 sample was found to be significantly different from the region represents.[1] Consequently, the samples were accepted as being representative of their respective universes.

The first part of the hypothesis was tested using three different sets of manuscript census data. In the first test, the 5229 farm 1860 manuscript census sample was divided into two parts--Old South and New South. The Old South sample consisted of farms selected from Virginia, North Carolina, South Carolina, Georgia and Florida and the New South sample contained farms from Alabama, Mississippi, Tennessee, Louisiana, Arkansas and Texas. Each of the samples was tested for representativeness and neither was rejected as being unrepresentative of its universe. The size distributions of farms and size distributions of slaveholdings in the two regions were then compared. Smaller farms and nonslaveholders were found to have been relatively more numerous in the New South than in the Old South. Further comparisons revealed that farmers who operated smaller farms owned a greater proportion of wealth and produced a greater percentage of crops in the New South than did their counterparts in the Old South. Also, nonslaveholders were relatively more significant owners of wealth and producers of crops in the New South than in the Old South. Thus, we found that the small farmers and nonslaveholders were more common in the New South than in the Old South and our hypothesis that the large planters led the migration into the New South appeared doubtful.

---

[1]Throughout the present chapter a five percent level of significance is employed. Thus, a significant difference is one in which the observed results could have obtained only 5 percent of the time as a result of chance alone. In many cases the actual percentage was less than .05 percent.

The second test of the hypothesis utilized a subsample of the 5229 farm 1860 sample. This subsample consisted of the 964 sample farms which were selected from Gray's seventy-four major cotton producing counties. This 964 farm subsample was divided into three parts which Gray designated as Old South, Alluvial and New South regions. The samples from each of these regions were tested and their biases noted. (These biases were considered in arriving at the conclusions which follow.) The size distributions of farms and slaveholdings in the two regions were then compared as in the previous test. Small farms were found to have been relatively more numerous in the New South than in the Old South. Small farms were also a greater proportion of Alluvial farms than of Old South farms although the differences between the Alluvial and Old South distributions were not so great as between New and Old South. Nonslaveholders were much more common in the New South than in either the Alluvial or Old South regions and, in fact, the Alluvial and Old South size distributions of slaveholdings were not significantly different. If nonslaveholders were omitted from the size distribution of slaveholdings, the New and Old South distributions were not significantly different but, in the Alluvial regions, slaveholdings appeared to have been larger than in the Old South. Further comparisons revealed that the smaller farmers in the New South owned a greater percentage of the regional wealth and produced a larger share of the crops than did their counterparts in the other two regions. Nonslaveholders accounted for a greater percentage of wealth owned and crops produced in the New South than in the other regions; in the Alluvial, nonslaveholders accounted for a smaller share. The share of wealth owned and crops produced by large slaveholders was greatest in the Alluvial region.

Again, the hypothesis that larger planters led the migration into the Southwest appeared questionable. However, we began to understand the source of our hypothesis.

The final test of part (a) of our hypothesis made use of the 897 farm 1850 sample. This sample also was divided into three parts--Old South, Alluvial, and New South--based on Gray's classifications. Again, the regional samples were tested and the biases noted. The size distributions of farms and size distributions of slaveholdings in the three regions were compared and the following differences noted. First, small farms and nonslaveholders were relatively more numerous in the New South than in the Old South region. Nonslaveholders were more common in the New South than in the Alluvial region but the size distributions of farms in the two regions were substantially the same. Large slaveholdings were more common in the Alluvial region than in either the Old or New South. Second, small farmers and small slaveholders in the New South accounted for a greater share of wealth owned and crops produced than did their counterparts in the Old South. In the Alluvial, small farmers produced approximately the same share of regional crop production as did their counterparts in the Old South but small _slaveholders_ produced a considerably smaller share of regional crop production in the Alluvial than in either the Old or New South. Thus, the 1850 data failed to support our hypothesis that large planters led the migration into the Southwest. Again, it appeared that the source of our hypothesis may have been an overemphasis on the fairly limited Alluvial region by past historians.

Thus, we tested the first part of our hypothesis using three different sets of manuscript census data. In each case, the data failed to support the hypothesis that large planters led the migration into the Southwest. However, it appeared that our hypothesis might be valid if applied only to the migration into the Alluvial regions.

Our test of the second part of the hypothesis utilized data for the Old South, Alluvial, and New South regions selected from both the 1850 and 1860 manuscript censuses. The size distributions of farms and slaveholdings in 1850 for each region were compared with the 1860 distributions which obtained in each region. The 1850 and 1860 size distributions of farms in the Old South region were found to have been significantly different due to the larger percentage of farms of one hundred acres or more in 1860. In the Alluvial and New South regions, the 1850-1860 differences were not statistically significant. However, a greater percentage of Alluvial farms contained one hundred acres or more in 1860 than in 1850. Thus, larger landowners became relatively more numerous in the Old South and, possibly, in the Alluvial during the decade of the 1850's. In the case of slaveholdings, the same observations obtained. The percentage of slaveholders with ten or more slaves increased in the Old South from 1850 to 1860 and, again, the increase was statistically significant. The percentage of Alluvial farms with twenty slaves or more increased between 1850 and 1860 but this increase was not statistically significant. In the New South, no significant changes in slaveholdings appeared. Further comparisons indicated that the share of Old South crops produced and wealth owned by the larger farmers and larger slaveholders increased during the 1850's. A similar increase occurred in the Alluvial region. In the New South, the share of crops produced and wealth owned by the large landowners and larger slaveholders declined between 1850 and 1860.

In the Alluvial regions, analysis of the values per acre of large and small farms indicated that the smaller farmer was being pushed off the better quality land between 1850 and 1860. In the Old South, the smaller farmer appears to have held land of better than average quality in both 1850 and 1860. Further, in 1850 land prices in the Old South exceeded those in the New South. Thus, any conclusion that smaller farmers were being pushed out of the Old South may indicate that they were being driven off the better quality land. By 1860, the value of New South land was greater than Old South land and the displacement of smaller farmers may have ceased.

Our examination of the role of the smaller farmer in primary cotton producing regions yielded the impression that the smaller farmer generally farmed land of inferior quality and held a larger share of the unimproved than the improved acreage. On this land, he raised relatively more food crops than his larger neighbors did and, consequently, relatively less cotton. He raised a relatively larger share of livestock, other than the sheep and mules, reflecting his larger unimproved acreage holdings. His animals were of lower quality than his larger neighbors' which may reflect a diet of mast rather than corn. Relative to all southerners, his own table was likely well set with milk, meat, peas, beans, and Irish potatoes for he raised these products in proportions greater than his share of the total population. His diet cannot be compared to that of other free southerners, however, without knowing how the larger planters distributed various foodstuffs between free persons and slaves. Cotton was the smaller farmer's cash crop but he may have supplemented his income with sales of foodstuffs and manufactured items.

CONCLUSIONS FROM THE STUDY

The upshot of this study is that the data from the 1850 and 1860 manuscript census samples clearly <u>fail</u> to support either portion of our hypothesis.

First, consider part (a): that the larger planters led, accompanied, or immediately followed the smaller farmer into the Southwest. Our comparison of the Old and New South regions in 1860 revealed that smaller farmers and nonslaveholders were relatively more numerous in the New South than in the Old South. Clearly, large planters did not lead the migration. In fact, their participation in the migration was certainly not in proportion to their numerical representation in the Old South. Had their participation been proportional to their numbers in the Old South, the Old and New South would have contained similar size distributions of farms and approximately the same percentage of nonslaveholders. Thus, the 1860 sample, selected from the 405 counties which produced at least one thousand bales of cotton in 1860, fails to support the first part of our hypothesis. Because this sample contains data from virtually every important cotton producing county, we can generalize to the cotton-producing South as a whole and suggest that westward migration in the cotton producing regions does not appear to have been a migration of large planters leading, accompanying, or closely following the small farmers. Rather, the migration appears to be predominantly one of small farmers and nonslaveholders.

If we test the first part of the hypothesis using data from Gray's seventy-four major cotton producing regions in 1860, the results are the same. Small farms were more common in the New South than in the Old South <u>and</u> relatively more numerous in the Alluvial than in the Old South.

Nonslaveholders were a greater percentage of farmers in the New South than in either of the other regions. Once again, the analysis fails to support our hypothesis. However, we did find that large slaveholdings appear to have been more common in the Alluvial than in the other regions and these large slaveholders produced a greater share of the Alluvial crops than did their counterparts in the other regions. It begins to appear that the source of our hypothesis may be an emphasis on the Alluvial region and a lack of attention to the other regions of the Southwest. This emphasis is easy to understand for the Alluvial region did contain both rich land and a great water transportation system. However, less than fourteen percent of the 1849 cotton crop and less than twenty percent of the 1859 crop was produced in this region.

Our third test of part (a) of our hypothesis utilized data from the 1850 manuscript census returns from Gray's major cotton producing counties. Again, the data fail to support our hypothesis that the larger planters led the migration into the Southwest. In 1850, small farms and nonslaveholders were relatively more numerous in the New South than in the Old South region. Large slaveholders, however, were more common in the Alluvial region than in either the Old or New South. Again it appears that an emphasis on the Alluvial is the source of our hypothesis.

Thus, using three different sets of manuscript data we arrive at the same conclusion: the data fail to support the notion that large planters led, accompanied, or immediately followed the smaller farmers into the Southwest. Larger planters may have been important factors in migration into the Alluvial but, even here, the results are not statistically significant.

Part (b) of our hypothesis fared little better. Comparisons of the 1850 and 1860 size distributions of farms and slaveholdings in the three regions indicated that the _only_ statistically significant changes which occurred during the 1850's took place in the Old South region. In this region, smaller farmers appeared to have been bought out by larger planters. In the Alluvial, also, the percentage of larger farmers and larger slaveholders appeared to increase during the decade but the 1850-1860 difference in this region was not statistically significant. In the New South, there is no evidence that the smaller farmers were being "pushed out" by larger planters. Thus, we conclude that in the Southwest as a whole the evidence fails to support the hypothesis that the small farmers were pushed out by larger planters who were able to buy the better land from them. In the Alluvial, some "pushing out" may have occurred.

The role of the small farmer in each of the regions is about what we had anticipated. He generally held land of somewhat inferior quality and held a larger share of the improved than the unimproved acreage. He raised relatively more food crops than his larger neighbors did and, consequently, relatively less cotton. Yet cotton was his cash crop. He raised a relatively larger proportion of animals, other than sheep and mules, reflecting his unimproved acreage holdings. However, he does not appear to have been a herder or merely a subsistence farmer raising only food crops. In fact, he was able to raise food crops in proportions greater than his share of the total population and have land and labor left over to devote to the production of cotton. However, we cannot support Robert E. Russel's statement that "perhaps one-half the cotton was grown on farms where there were either no slaves at all or fewer than ten or a dozen."[2] In fact, in the Old and New South regions only 17.5 percent of the cotton was produced on farms with less than ten slaves. In the Alluvial regions this percentage was 7.7 percent.

The conclusions above are subject to two modifications. First, the conclusions are based on manuscript census samples selected from cotton producing regions. Consequently, the conclusions are not necessarily applicable to regions where rice, sugar, or tobacco were the important crops. Second, the data are selected from the 1850 and 1860 censuses. The study covers only a single decade at the end of the antebellum period. Consequently, the conclusions, strictly speaking, apply only to that period. However, it is the writer's contention that this decade is the most important one to test. In the first place, migration was clearly taking place during the period. Further, because this decade is the final one of the antebellum period any large planters who took part in the migration into the Southwest had done so by 1860. Thus, we have given the large planters every chance to move, looked for him in the newer regions, and found him missing.

IMPLICATIONS OF THIS STUDY

Having tested our hypothesis and examined the role of the smaller farmer in each of the regions, three generalizations seem appropriate. First, southern migration was a movement of the smaller farmers and nonslaveholders in proportions greater than their representation in the

---

[2] "The Effects of Slavery upon Nonslaveholders in the Ante-Bellum South," Agricultural History, XV (April, 1941) reprinted in Harold P. Woodman, Slavery and the Southern Economy (New York: Harcourt, Brace & World, 1966), p. 120.

older regions. Second, if smaller farmers were forced to sell out the best land to the larger planters, the selling took place largely in the Old South and in the Alluvial regions. Third, because these smaller farmers, especially in the newer regions, were frequently nonslaveholders and because one-fifth of the improved land and one-fourth of the unimproved land in the Southwest was in the hands of nonslaveholders in 1860, an immediate need for more slave territory is not apparent. In fact, it appears that southern demands for increased slave territory were made for political or long-range expansion purposes rather than to provide for the immediate expansion of plantation agriculture.

CONTRIBUTIONS OF THIS STUDY

Perhaps it is appropriate at this point to briefly comment on the contributions this study has hoped to make. First and most obvious, the study has attempted to cast more light upon the subject of westward migration in the antebellum South. Thus, the two-part hypothesis was tested and found wanting.

Second, the data employed in testing the hypothesis are new. To the writer's knowledge, the 1850 and 1860 manuscript censuses have never been so extensively mined. Further, the techniques employed in the mining are new. The samples are large, extensive and have been tested to determine their representativeness, i.e., their right to "stand for" the regions they presume to represent. Finally, care has been taken to determine whether the observed differences are statistically significant --a must when sample data are employed.

Third, several of the size distributions utilized in the study have not previously been available. For example, size distributions of 1850 farms by region are not available in the published volumes. (In 1860, size distributions of farms by county were included in the published census. For 1850, county data are not available.) Also, in 1850, size distributions of slaveholdings were not included in the published census. Further, size distributions of <u>rural</u> slaveholdings have not previously been available for either 1850 or 1860.

Finally, the study focuses attention on the role of the smaller farmers and nonslaveholders using the very extensive record left us in the manuscript census. The relative contributions of the various farm size and slaveholding classes to the economies of three cotton producing regions have been examined. Prior to the selection of the two manuscript census samples used in this study, such an examination was hardly possible.

ADDENDUM

Having discarded the traditional view of westward migration in the antebellum South, something should be offered in its place. Perhaps it is sufficient to suggest that more attention should be paid to the role of the little man, the plain folk rather than the "heroes" in history who, by the sheer weight of their numbers, are heavy contributors to many historical movements. Truly, God must have liked them for He created so many.

APPENDIX I

SIZE AND SCOPE OF THE MATCHED 1860 SAMPLE

The purpose of this series of studies is to focus attention upon the organization of agricultural production within a region producing a large export crop--cotton. The basis of the studies is a matched sample of farms selected from the 1860 manuscript census returns of the "cotton counties" (counties producing 1,000 or more 400 pound bales of cotton in 1860).

The total sample includes data for 5229 farms in eleven southern states. Of the 409 "cotton counties," 382 were represented in the sample. Four of the omitted counties were in Missouri for which data was not available; five were too small to be included in the sample; fifteen had incomplete census returns; three were omitted due to matching difficulties.

Incomplete data, matching difficulties and other problems reduced the sample size by 426 farms (eight per cent of the planned sample).

Several farms listed in the manuscript census appear to have been operated in conjunction with another farm or farms. (For example, several farms listed separately in the agricultural census were owned by members of the same household.) The sample farms include 143 farms formed by combining the crop and population data from two or more related farms.

The size and scope of the sample, by state, are summarized on the following pages.

SUMMARY
Eleven States

Cotton Production: 5,387,052 bales

Cotton Producing States: 13

States Sampled: 11 (producing 5,344,166 bales, 99% of U.S. total)

Alabama	North Carolina
Arkansas	South Carolina
Florida	Tennessee
Georgia	Texas
Louisiana	Virginia
Mississippi	

States Omitted: Illinois - No counties produced 1,000 bales.
Missouri - Manuscript census microfilm unavailable.

Cotton Counties: 409

Counties Sampled: 382 (producing 4,994,940 bales, 93% of U.S. total)

Counties Omitted:

14	- Incomplete agricultural census
5	- Too few farms
4	- Missouri microfilm unavailable
3	- Matching difficulties
1	- Poor quality microfilm
27	Total

Sample Size: 5229 farms

Farms Omitted:

195	- Due to omitting twenty-seven counties
129	- Due to matching difficulties
99	- Due to incomplete data
3	- Other
426	Total (8% of planned sample)

Farms Combined: The sample farms include 143 farms formed by combining the crop and population data from two or more related farms.

Alabama

Cotton Production: 989,955 bales - 18.4% of U.S. total

Total Counties: 52

Cotton Counties: 50

Counties Sampled: 49 (producing 986,991 bales, 99% of Alabama total)

Autauga	Covington	Lowndes	Randolph
Barbour	Dale	Madison	Russell
Bibb	Dallas	Marengo	Shelby
Blount	De Kalb	Marion	St. Claire
Butler	Fayette	Marshall	Sumter
Calhoun	Franklin	Macon	Tallapossa
Chambers	Green	Montgomery	Talladega
Cherokee	Henry	Monroe	Tuscaloosa
Choctaw	Jackson	Morgan	Walker
Clarke	Jefferson	Perry	Washington
Coffee	Lawrence	Pickens	Wilcox
Conecuh	Lauderdale	Pike	
Coosa	Limestone		

Cotton Counties Omitted: 1
Baldwin - Too few farms (118)

Sample Size: 1005 farms. Approximately 18% of total sample.

Farms Omitted: 20 due to incomplete data
5 due to matching difficulties
25 total (3% of planned sample)

Farms Combined: Twenty-seven sample farms are the result of combining two farms.

Arkansas

Cotton Production: 367,393 bales - 6.8% of U.S. total

Total Counties: 55

Cotton Counties: 36

Counties Sampled: 33 (producing 301,153 bales, 82% of Arkansas total)

Arkansas	Franklin	Phillips
Ashley	Hempstead	Poinsett
Bradley	Hot Spring	Pope
Calhoun	Independence	Prairie
Clark	Jefferson	Pulaski
Columbia	Johnson	Saline
Conway	Lafayette	St. Francis
Crittenden	Marion	Serier
Dallas	Mississippi	Union
Desha	Monroe	White
Drew	Perry	Yell

Cotton Counties Omitted: 3
Chicot - Incomplete agricultural census
Jackson - Incomplete agricultural census
Ouachita - Incomplete agricultural census

Sample Size: 434 farms. Approximately 9% of total sample.

Farms Omitted: 45 due to omitting three entire counties
21 due to matching difficulties
66 total (13% of planned sample)

Farms Combined: Eleven of the sample farms are the result of combining two or more farms.

Florida

Cotton Production: 65,153 bales - 1.2% of U.S. total

Total Counties: 37

Cotton Counties: 9

Counties Sampled: 9 (producing 57,565 bales, 88% of Florida total)

Alachua	Hamilton	Leon
Columbia	Jackson	Madison
Gadsden	Jefferson	Marion

Cotton Counties Omitted: None

Sample Size: 65 farms. Approximately 1% of total sample.

Farms Omitted: None

Farms Combined: One of the sample farms is the result of combining two farms.

Georgia

Cotton Production: 701,840 bales - 13.0% of U.S. total

Total Counties: 132

Cotton Counties: 86

Counties Sampled: 82 (producing 646,554 bales, 92% of Georgia total)

Baker	Crawford	Jasper	Poulding	Troup
Baldwin	Decatur	Jefferson	Pike	Twiggs
Bibb	De Kalb	Johnson	Polk	Upson
Bullock	Dooly	Jones	Pulaski	Walton
Burke	Elbert	Laurens	Putnam	Warren
Butts	Emanuel	Lee	Quitman	Washington
Calhoun	Fayette	Liberty	Randolph	Webster
Campbell	Floyd	Lincoln	Richmond	Wilkes
Carroll	Glasscock	Lowndes	Schley	Wilkinson
Cass	Green	Macon	Scriven	Worth
Chattahoochee	Gwinnett	Madison	Spalding	
Chattooga	Hancock	Marion	Stewart	
Clark	Hart	Meriwether	Sumter	
Clay	Harris	Monroe	Talbot	
Clayton	Heard	Morgan	Taliaferro	
Cobb	Henry	Muscogee	Taylor	
Columbia	Houston	Newton	Terrell	
Coweta	Jackson	Oglethorpe	Thomas	

Cotton Counties Omitted: 4
- Brooks - Poor quality of agricultural census.
- Dougherty - Difficulty matching free population.
- Early - Poor quality of agricultural census.
- Mitchell - Difficulty matching free population.

Sample Size: 817 farms. Approximately 15% of total sample.

Farms Omitted:

20	due to omitting four entire counties
10	due to incomplete data
7	due to matching difficulties
1	due to accidental omission
38	total (4% of planned sample)

Farms Combined: Sixteen of the sample farms are the result of combining two farms.

Louisiana

Cotton Production: 777,738 bales - 14.4% of U.S. total

Total Counties: 48

Cotton Counties: 32

Counties Sampled: 27 (Producing 712,765 bales, 92% of Louisiana total)

Avoyelles	Felicinia, W.	Rapides
Baton Rouge, E.	Franklin	Sabine
Bossier	Jackson	St. Helena
Carroll	Lafayette	St. Landry
Catahoula	Livingston	St. Martin
Claiborne	Madison	Tensas
Concordia	Morehouse	Union
De Soto	Natchitoches	Washington
Felicinia, E.	Quichita	Winn

Cotton Counties Omitted: 5

- Baton Rouge, W. - Incomplete agricultural census.
- Cadds - Incomplete agricultural census.
- Caldwell - Incomplete agricultural census.
- Point Coupee - Slave census microfilm poor.
- Vermillion - Agricultural census appears falsified.

Sample Size: 226 farms. Approximately 5% of total sample.

Farms Omitted:
35 due to omitting five entire counties.
10 due to matching difficulties.
4 due to incomplete data.
49 total (18% of planned sample).

Farms Combined: Eight of the sample farms are the result of combining two or more farms.

Mississippi

Cotton Production: 1,202,507 bales - 22.3% of U.S. total

Total Counties: 60

Cotton Counties: 53

Counties Sampled: 48 (producing 1,099,477 bales, 92% of Mississippi total)

Adams	De Soto	Lowndes	Rankin
Amite	Franklin	Madison	Scott
Attala	Hinds	Marion	Simpson
Boliver	Holmes	Marshall	Smith
Calhoun	Itawamba	Monroe	Tallahatchie
Carroll	Jasper	Neshoba	Tippah
Chickasaw	Jefferson	Newton	Tishomingo
Choctaw	Kemper	Noxubee	Warren
Claiborne	Lafayette	Oktibbeha	Wayne
Clark	Lauderdale	Panola	Wilkinson
Copiah	Lawrence	Pike	Winston
Covington	Leake	Pontotoc	Yazoo

Cotton Counties Omitted: 5
- Coahoma - Incomplete agricultural census.
- Harrison - Too few farms (60).
- Issaquena - Too few farms (104).
- Tunica - Too few farms (112).
- Yalabusha - Incomplete and poor quality agricultural census.

Sample Size: 707 farms. Approximately 14% of total sample.

Farms Omitted: 20 due to omitting five entire counties
23 due to matching difficulties
43 total (6% of planned sample).

Farms Combined: Twenty-four of the sample farms are the result of combining two or more farms.

North Carolina

Cotton Production: 145,514 bales - 2.7% of U.S. total

Total Counties: 86

Cotton Counties: 26

Counties Sampled: 24 (producing 123,401 bales, 85% of North Carolina Total)

Anson	Johnston	Richmond
Caberras	Jones	Robeson
Duplin	Lenoir	Rowan
Edgecomb	Martin	Surry
Franklin	Mecklenburg	Union
Green	Nash	Wake
Halifax	Northampton	Wayne
Hertford	Pitt	Wilson

Cotton Counties Omitted: 2
Bertie - Incomplete agricultural census.
Montgomery - Incomplete agricultural census.

Sample Size: 393 farms. Approximately 8% of total sample.

Farms Omitted:
30 due to omitting two entire counties.
20 due to matching difficulties.
1 due to incomplete data.
1 other.
52 Total (12% of planned sample).

Farms Combined: Eight of the sample farms are the result of combining two or more farms.

South Carolina

Cotton Production: 353,412 bales - 6.6% of U.S. total

Total Counties: 30

Cotton Counties: 27

Counties Sampled: 27 (producing 351,920 bales, 99% of South Carolina total)

Abbeville	Darlington	Mailborough
Anderson	Edgefield	Newberry
Barnwell	Fairfield	Orangebush
Beaufort	Greenville	Richland
Charleston	Kershaw	Spartanburgh
Chester	Lancaster	Sumter
Chesterfield	Laurens	Union
Clarendon	Lexington	Williamsburgh
Colleton	Marion	York

Cotton Counties Omitted: None

Sample Size: 524 farms. Approximately 10% of total sample.

Farms Omitted: 27 due to incomplete data.
9 due to matching difficulties.
36 total (8% of planned sample).

Farms Combined: Ten of the sample farms are the result of combining two farms.

Tennessee

Cotton Production: 296,464 bales - 5.5% of U.S. total

Total Counties: 84

Cotton Counties: 19

Counties Sampled: 19 (producing 287,652 bales, 97% of Tennessee total)

Carroll	Lauderdale	Maury
Fayette	Lawrence	Monroe
Gibson	Lincoln	Rutherford
Giles	Macon	Shelby
Hardeman	McNairy	Tipton
Haywood	Madison	Williamson
Henderson		

Cotton Counties Omitted: None

Sample Size: 465 farms. Approximately 8% of total sample.

Farms Omitted: 32 due to incomplete data.
22 due to matching difficulties.
1 other
55 total (11% of planned sample).

Farms Combined: Twenty-four of the sample farms are the result of combining two farms.

Texas

Cotton Production: 431,463 bales - 8.0% of U.S. total

Total Counties: 151

Cotton Counties: 62

Counties Sampled: 59 (Producing 416,716 bales, 97% of Texas total)

Anderson	Fayette	Liberty	Sabine
Angelina	Fort Bend	Limestone	San Augustine
Austin	Freestone	Madison	Shelby
Bastrop	Golaid	Matagorda	Smith
Bowie	Gonzales	McLennan	Titus
Brazoria	Grimes	Milan	Travis
Burleson	Guadalupe	Montgomery	Trinity
Caldwell	Harrison	Macogdoches	Tyler
Cass	Hayes	Navarro	Upshur
Cherokee	Henderson	Newton	Victoria
Colorado	Houston	Panola	Walker
Comal	Jasper	Polk	Washington
DeWitt	Lamar	Red River	Wharton
Falls	Lavaca	Robertson	Wood
Fannin	Leon	Rusk	

Cotton Counties Omitted: 3
Brazos - Incomplete agricultural census and matching difficulties.
Jackson - Difficulty matching slave census.
Marion - Too few farms (141).

Sample Size: 551 farms. Approximately 10% of total sample.

Farms Omitted: 10 due to omitting two entire counties.
9 due to matching difficulties.
5 due to incomplete data.
24 total (4% of planned sample).

Farms Combined: Fourteen of the sample farms are the result of combining two farms.

Virginia

Cotton Production: 12,727 bales - 0.2% of U.S. total

Total Counties: 148

Cotton Counties: 5

Counties Sampled: 5 (producing 10,746 bales, 84% of Virginia total)

Essex
Greenville
Prince George
Southampton
Sussex

Cotton Counties Omitted: None

Sample Size: 42 farms. Approximately 1% of total sample.

Farms Omitted: 3 due to matching difficulties (7% of the planned sample).

Farms Combined: None.

APPENDIX II

Data Tape Format - 1860 Sample

Word No.	Contents
1	County No.
2	Soil type code
3	No. Farms sampled
4	County cotton production (100's of bales)
5	Page-line code
6	Val. pers. prop.
7	Acres of improved land
8	Acres of unimproved land
9	$ Val. of farm
10	$ Val. of farm machinery
11	Head of horses
12	Head of asses & mules
13	Head of milk cows
14	Head of working oxen
15	Head of other cattle
16	Head of sheep
17	Head of swine
18	$ Val. of livestock
19	Bushels of wheat
20	Bushels of rye
21	Bushels of corn
22	Bushels of oats
23	Pounds of rice
24	Pounds of tobacco
25	Bales of cotton (400# bales)
26	Pounds of wool
27	Bushels of peas & beans
28	Bushels of Irish potatoes
29	Bushels of sweet potatoes
30	Bushels of barley
31	Bushels of buckwheat
32	$ Val. of orchard production
33	Gal. of wine
34	$ Val. of mkt. garden production
35	Pounds of butter
36	Pounds of cheese
37	Tons hay & Fodder
38	Bushels clover seed
39	Bushels grass seed
40	Pounds hops
41	Tons of dew rotted hemp
42	Tons of water rotted hemp
43	Tons of other hemp
44	Pounds of flax
45	Bushels flaxseed

Word No.	Contents	
46	Pounds of silk cocoons	
47	Pounds of maple sugar	
48	Cane sugar (1000 pounds)	
49	Gal. molasses	
50	Pounds of beeswax	
51	Pounds of honey	
52	$ Val. home manufactures	
53	$ Val. animals slaughtered	
54	Check total for above 53 items	
55	Free males less than 6 mo.	
56	Free males	1 - 4
57	Free males	5 - 9
58	Free males	10 - 14
59	Free males	15 - 19
60	Free males	20- 24
61	Free males	25 - 29
62	Free males	30 - 34
63	Free males	35 - 39
64	Free males	40 - 44
65	Free males	45 - 49
66	Free males	50 - 54
67	Free males	55 - 59
68	Free males	60 - 64
69	Free males	65 - 69
70	Free males	70 - 79
71	Free males	80 - 89
72	Free males	90 & over
73	Free females	less than 6 mo.
74	Free females	1 - 4
75	Free females	5 - 9
76	Free females	10 - 14
77	Free females	15 - 19
78	Free females	20 - 24
79	Free females	25 - 29
80	Free females	30 - 34
81	Free females	35 - 39
82	Free females	40 - 44
83	Free females	45 - 49
84	Free females	50 - 54
85	Free females	55 - 59
86	Free females	60 - 64
87	Free females	65 - 69
88	Free females	70 - 79
89	Free females	80 - 89
90	Free females	90 & over
91	Slave male	less than 6 mo.
92	Slave male	1 - 4
93	Slave male	5 - 9
94	Slave male	10 - 14
95	Slave male	15 - 19
96	Slave male	20 - 24

Word No.	Contents	
97	Slave male	25 - 29
98	Slave male	30 - 34
99	Slave male	35 - 39
100	Slave male	40 - 44
101	Slave male	45 - 49
102	Slave male	50 - 54
103	Slave male	55 - 59
104	Slave male	60 - 64
105	Slave male	65 - 69
106	Slave male	70 - 79
107	Slave male	80 - 89
108	Slave male	90 & over
109	Slave female	less than 6 mo.
110	Slave female	1 - 4
111	Slave female	5 - 9
112	Slave female	10 - 14
113	Slave female	15 - 19
114	Slave female	20 - 24
115	Slave female	25 - 29
116	Slave female	30 - 34
117	Slave female	35 - 39
118	Slave female	40 - 44
119	Slave female	45 - 49
120	Slave female	50 - 54
121	Slave female	55 - 59
122	Slave female	60 - 64
123	Slave female	65 - 69
124	Slave female	70 - 79
125	Slave female	80 - 89
126	Slave female	90 & over
127	Farm laborer male	less than 6 mo.
128	Farm laborer male	1 - 4
129	Farm laborer male	5 - 9
130	Farm laborer male	10 - 14
131	Farm laborer male	15 - 19
132	Farm laborer male	20 - 24
133	Farm laborer male	25 - 29
134	Farm laborer male	30 - 34
135	Farm laborer male	35 - 39
136	Farm laborer male	40 - 44
137	Farm laborer male	45 - 49
138	Farm laborer male	50 - 54
139	Farm laborer male	55 - 59
140	Farm laborer male	60 - 64
141	Farm laborer male	65 - 69
142	Farm laborer male	70 - 79
143	Farm laborer male	80 - 89
144	Farm laborer male	90 & over

Word No.	Contents	
145	Farm laborer female	less than 6 mo.
146	Farm laborer female	1 - 4
147	Farm laborer female	5 - 9
148	Farm laborer female	10 - 14
149	Farm laborer female	15 - 19
150	Farm laborer female	20 - 24
151	Farm laborer female	25 - 29
152	Farm laborer female	30 - 34
153	Farm laborer female	35 - 39
154	Farm laborer female	40 - 44
155	Farm laborer female	45 - 49
156	Farm laborer female	50 - 54
157	Farm laborer female	55 - 59
158	Farm laborer female	60 - 64
159	Farm laborer female	65 - 69
160	Farm laborer female	70 - 79
161	Farm laborer female	80 - 89
162	Farm laborer female	90 & over
163	Overseers male	less than 6 mo.
164	Overseers male	1 - 4
165	Overseers male	5 - 9
166	Overseers male	10 - 14
167	Overseers male	15 - 19
168	Overseers male	20 - 24
169	Overseers male	25 - 29
170	Overseers male	30 - 34
171	Overseers male	35 - 39
172	Overseers male	40 - 44
173	Overseers male	45 - 49
174	Overseers male	50 - 54
175	Overseers male	55 - 59
176	Overseers male	60 - 64
177	Overseers male	65 - 69
178	Overseers male	70 - 79
179	Overseers male	80 - 89
180	Overseers male	90 & over
181	Overseers female	less than 6 mo.
182	Overseers female	1 - 4
183	Overseers female	5 - 9
184	Overseers female	10 - 14
185	Overseers female	15 - 19
186	Overseers female	20 - 24
187	Overseers female	25 - 29
188	Overseers female	30 - 34
189	Overseers female	35 - 39
190	Overseers female	40 - 44
191	Overseers female	45 - 49
192	Overseers female	50 - 54
193	Overseers female	55 - 59
194	Overseers female	60 - 64

Word No.	Contents	
195	Overseers female	65 - 69
196	Overseers female	70 - 79
197	Overseers female	80 - 89
198	Overseers female	90 & over
199	Non-farmworkers male	less than 6 mo.
200	Non-farmworkers male	1 - 4
201	Non-farmworkers male	5 - 9
202	Non-farmworkers male	10 - 14
203	Non-farmworkers male	15 - 19
204	Non-farmworkers male	20 - 24
205	Non-farmworkers male	25 - 29
206	Non-farmworkers male	30 - 34
207	Non-farmworkers male	35 - 39
208	Non-farmworkers male	40 - 44
209	Non-farmworkers male	45 - 49
210	Non-farmworkers male	50 - 54
211	Non-farmworkers male	55 - 59
212	Non-farmworkers male	60 - 64
213	Non-farmworkers male	65 - 69
214	Non-farmworkers male	70 - 79
215	Non-farmworkers male	80 - 89
216	Non-farmworkers male	90 & over
217	Non-farmworkers female	less than 6 mo.
218	Non-farmworkers female	1 - 4
219	Non-farmworkers female	5 - 9
220	Non-farmworkers female	10 - 14
221	Non-farmworkers female	15 - 19
222	Non-farmworkers female	20 - 24
223	Non-farmworkers female	25 - 29
224	Non-farmworkers female	30 - 34
225	Non-farmworkers female	35 - 39
226	Non-farmworkers female	40 - 44
227	Non-farmworkers female	45 - 49
228	Non-farmworkers female	50 - 54
229	Non-farmworkers female	55 - 59
230	Non-farmworkers female	60 - 64
231	Non-farmworkers female	65 - 69
232	Non-farmworkers female	70 - 79
233	Non-farmworkers female	80 - 89
234	Non-farmworkers female	90 & over
235	Total Population	

APPENDIX III

ACREAGE, VALUE AND CROP OUTPUTS OF SAMPLE FARMS AS A PERCENTAGE OF COTTON COUNTY TOTALS--1860

	ALA.	ARK.	FLA.	GA.	LA.	MISS.	N.C.	S.C.	TENN.	TEX.	VA.
No. of Farms											
In State[a]	54,342	25,593	3,331	44,480	14,021	41,954	23,788	30,582	24,925	31,097	2,325
In Sample	1,005	434	65	817	226	707	393	524	465	551	42
% Sampled	1.93	1.70	1.95	1.84	1.61	1.72	1.65	1.72	1.87	1.77	1.81
Improved Acres	1.91	1.78	1.29	2.78	2.00	1.79	1.45	1.97	2.21	1.91	1.71
Unimproved Acres	1.96	1.81	.82	2.03	2.45	1.74	2.14	1.77	2.35	1.78	1.16
Cash Val. of Farm	1.85	1.68	1.10	2.18	2.90	1.67	1.73	1.72	2.55	2.12	1.74
Val. of Imp. & Mach.	1.91	1.66	1.00	2.15	1.93	1.75	1.55	1.72	2.23	2.46	1.83
Horses	1.93	1.84	1.44	1.98	1.64	1.83	1.84	1.86	2.02	2.32	1.92
Asses & Mules	1.80	2.16	1.17	2.17	2.19	1.69	1.39	1.73	1.95	2.23	1.53
Milch Cows	1.94	2.02	1.72	1.85	1.69	1.78	1.80	1.82	1.95	1.72	1.88
Working Oxen	2.01	2.12	1.24	2.01	1.72	1.67	1.76	1.56	2.12	2.11	2.05
Other Cattle	1.85	1.86	1.56	1.83	1.73	1.63	1.67	1.88	1.88	2.18	1.88
Sheep	1.96	2.36	1.62	1.95	1.63	1.72	1.73	2.05	2.16	3.33	2.08
Swine	1.95	1.81	1.67	2.04	1.92	1.72	1.83	1.89	2.00	2.02	1.51
Val. of Livestock	1.84	2.07	1.29	1.96	1.95	1.75	1.83	1.86	1.99	2.14	1.63
Wheat	1.91	1.75	1.10	2.65	.48	2.06	1.31	1.94	2.05	2.37	1.88
Rye	1.85	8.20	.18	1.71	.33	1.75	1.21	1.37	2.01	1.47	1.68
Corn	1.95	1.81	1.27	2.01	1.85	1.83	1.60	1.92	2.22	2.08	1.51
Oats	2.26	1.79	.13	2.66	1.29	1.72	1.38	1.84	2.37	2.68	1.20
Rice	14.95	*	1.21	3.11	9.12	2.52	2.50	.10	*	.08	*
Tobacco	1.30	4.53	.63	.27	.36	.28	1.22	.73	1.87	3.26	1.40
Cotton	1.74	1.46	.96	2.07	2.07	1.57	1.32	1.85	1.71	2.18	4.93
Wool	1.61	2.60	.61	1.80	1.82	1.72	1.60	2.13	2.43	3.65	2.61
Peas & Beans	2.14	2.41	2.41	2.34	1.85	2.00	1.64	1.70	1.93	1.77	1.25
Irish Potatoes	1.45	1.85	.92	2.42	2.25	1.94	2.15	1.92	1.69	2.40	1.80
Sweet Potatoes	2.12	2.07	1.56	2.27	2.38	1.68	1.91	1.66	2.02	2.13	1.95

	ALA.	ARK.	FLA.	GA.	LA.	MISS.	N.C.	S.C.	TENN.	TEX.	VA.
Val. of Orchard Products	2.19	2.91	.31	2.16	8.47	1.28	2.06	3.36	2.00	6.07	.89
Val. of Mkt. Garden Products	1.37	2.90	*	1.30	.28	1.56	2.44	1.66	1.78	5.07	*
Butter	1.93	2.02	1.69	2.02	2.20	1.91	1.63	1.79	2.25	1.96	2.03
Cheese	*	1.40	*	4.08	5.18	*	2.27	2.08	.59	1.45	*
Hay	1.77	1.54	1.46	3.76	2.02	21.87	1.71	1.46	1.87	2.17	1.25
Cane Sugar	*	*	2.65	24.85	2.28	*	*	*	*	2.96	*
Molasses	3.04	3.61	3.12	1.75	2.23	1.47	4.13	.67	1.76	5.90	*
Beeswax	1.53	2.65	1.65	2.53	1.51	1.71	1.91	2.10	2.00	2.95	1.52
Honey	1.48	2.14	1.30	2.21	1.81	1.72	1.83	2.13	1.66	2.39	4.44
Val. of Home Manuf.	1.89	2.01	2.09	2.25	2.01	1.79	1.69	1.75	2.15	1.91	2.44
Val. of Animals Slaughtered	1.92	2.04	1.47	2.28	2.07	1.89	1.80	1.84	2.12	2.01	1.79

*Not produced on farms included in the sample.

[a]Total Farms 3 acres and up in Cotton Counties times an adjustment factor. The adjustment factor for each state equals:

$$\frac{\text{Total Farms in State}}{\text{Farms 3 acres \& up in State}}$$

(Data on number of farms are from U.S. Bureau of the Census, _Eighth Census of the United States: 1860, Agriculture_.)

APPENDIX IV

VALUES OF t-STATISTICS BY STATE -- 1860

	ALABAMA		ARKANSAS		FLORIDA		GEORGIA		LOUISIANA		MISSISSIPPI		NORTH CAROLINA		SOUTH CAROLINA		TENNESSEE		TEXAS		VIRGINIA	
	cc[a]	sc[b]	cc	sc	cc	sc	cc	sc	cc	sc	cc	sc	cc	sc	cc	sc	cc	sc	cc	sc	cc	sc
Acres (improved)	.74	.74	.75	1.11	-2.91	same	1.49	1.52	1.77	2.07	1.56	1.51	-2.15	-.31	-2.14	same	1.77	same	1.00	1.10	-.22	same
Acres (unimproved)	1.33	1.34	.86	.81	-8.33		1.49	1.75	1.75	1.71	.99	1.19	1.12	1.14	.65		1.42		.02	.24	-2.77	
Cash Value of Farm	.06	.04	-.08	.09	-4.20		2.45	2.84	2.59	2.67	.53	1.22	.58	.17	.25		1.25		1.74	1.88	-.11	
Value of Implements and Machinery	.61	.59	-.22	.16	-5.74		2.21	2.68	.65	1.44	.67	.95	-.83	-1.18	.25		1.08		1.42	1.43	.03	
Horses	1.34	1.34	.40	.17	-3.30		2.20	2.20	.19	.76	2.70	2.37	2.59	2.33	2.60		1.55		1.19	1.20	.33	
Asses & Mules	-.27	-.27	.46	.47	-3.22		2.40	2.66	1.84	1.92	.40	.60	-1.56	-1.64	.36		.48		2.05	2.10	-.52	
Milch Cows	1.43	1.80	1.77	1.63	-.94		.30	.85	.42	.58	1.87	1.63	1.83	1.91	1.05		.81		-.55	-.63	.23	
Working Oxen	2.11	2.11	1.18	1.07	-2.13		1.98	1.98	.62	.26	.24	.25	.80	1.04	-.79		1.64		2.17	2.18	.41	
Other Cattle	.05	.45	1.10	1.00	-1.78		-.02	.90	.70	1.02	-.29	-.06	.25	2.02	1.67		.13		1.01	1.15	.26	
Sheep	1.01	1.10	1.57	1.50	-.49		.64	.77	.06	-.13	.59	.38	.56	.97	1.52		1.87		2.13	2.10	.44	
Swine	1.53	1.56	1.35	1.03	-1.75		2.20	2.50	1.41	.99	1.00	.82	1.79	1.78	1.98		.92		2.13	2.25	-1.31	
Value of Livestock	-.06	.00	1.69	1.74	-4.92		1.61	2.10	1.66	1.95	1.27	1.26	.81	.73	1.76		.81		2.03	2.17	-.47	
Wheat	.49	.44	.27	-.40	-.75		3.23	3.06	-3.50	-3.83	2.17	1.83	-2.51	-2.52	2.06		1.07		1.57	1.50	.10	
Rye	.04	.04	2.31	2.27	-15.67		-.23	-.16	-8.16	-9.04	.24	.07	-1.83	-1.96	-1.11		.29		-.84	-.79	-.11	
Corn	.76	.76	.99	.95	-3.14		1.93	2.21	.99	.18	1.83	1.78	-.49	-.39	1.84		2.48		2.20	2.22	-.91	
Oats	.92	1.15	.19	.08	-13.72		2.52	2.44	-.63	-.70	.09	.08	-1.05	-1.00	.49		1.26		.84	.83	-.94	
Rice	.95	.95	*	*	-.80		**	**	1.21	1.19	.54	.70	.60	.51	**		*		-20.14	-20.14	*	
Tobacco	-2.21	-2.21	.95	.89	-3.34		-12.53	-12.89	-4.63	-5.26	-12.09	-12.86	-.97	-1.24	-4.90		.03		.57	.56	-.82	
Cotton	-.77	-.79	-1.37	-.53	-4.53		1.47	1.81	1.42	1.25	-.73	-.26	-1.85	-1.97	.87		-.63		1.50	1.51	.83	
Wool	-1.78	-1.70	1.63	1.45	-3.58		-.14	.02	.55	.25	.44	.20	-.30	.54	1.33		2.34		1.65	1.65	.72	
Peas & Beans	1.76	1.74	1.74	1.49	.61		1.22	1.33	.61	.67	1.81	1.65	-.06	.09	.06		.19		.01	.21	-1.73	
Irish Potatoes	-3.45	-3.45	.90	.52	-3.31		1.73	1.66	1.32	2.07	1.95	1.74	1.38	1.46	.87		-.99		1.56	1.59	*	
Sweet Potatoes	1.45	1.47	2.40	2.03	-2.40		3.06	3.33	1.58	1.63	.31	-.13	1.87	1.61	-.21		.99		1.44	1.50	.42	
Value of Orchard Products	.76	.82	1.36	1.25	-7.44		.88	.77	.91	1.00	-1.44	-1.62	.87	.78	1.34		.34		1.27	1.28	*	
Value of Market Garden Products	-.67	-.27	.95	.84	*		-1.44	-1.58	-6.79	-2.45	-.16	-.31	.78	.67	-.04		-.09		1.78	1.77	*	
Butter	1.15	1.14	2.52	1.56	-.75		1.83	1.77	2.28	2.01	3.28	2.68	-.14	-.35	1.17		2.14		1.70	1.48	.38	
Cheese	*	*	-.22	-.30	*		.96	.96	.81	.77	*	*	.68	.63	.19		-2.94		-.36	-.39	*	
Hay	-.38	-.30	.30	-.61	1.46		**	**	.59	1.05	1.48	1.47	.29	.29	-1.73		.01		.78	.78	-2.76	
Cane Sugar	*	*	*	*	**		**	**	.46	.79	*	*	*	*	*		*		.62	.61	*	
Molasses	1.81	1.80	1.23	1.13	.69		-.22	-.53	.44	.83	-.25	-.38	1.12	1.15	-3.63		-.29		1.14	1.13	*	
Beeswax	-1.70	-1.70	1.70	1.44	-.18		2.29	2.29	-.17	-.47	.16	.02	1.03	1.21	1.27		.31		1.58	1.55	-.31	
Honey	-2.46	-2.44	1.35	.89	-.61		2.03	1.90	.52	.04	.37	.01	.74	.78	1.69		.87		1.30	1.24	.83	
Value of Home Manufactures	.48	.42	1.86	.92	.23		2.17	2.12	1.03	.78	-3.04	.47	.32	.61	.44		1.72		.73	.61	.64	
Value of Animals Slaughtered	.85	.86	2.69	1.81	-3.88		1.19	1.24	2.31	1.78	2.39	1.98	1.44	1.20	1.51		1.49		1.91	1.74	-.04	

* Not produced on farms included in the sample.
** Not available at time statistics were computed.
a Cotton Counties
b Sampled Counties

APPENDIX V

VALUES OF CHI-SQUARE STATISTICS FOR VARIOUS TESTS OF THE 1860 SAMPLE
(APPROXIMATE PROBABILITY LEVELS IN PARENTHESES)

	ALABAMA		ARKANSAS		FLORIDA		GEORGIA		LOUISIANA		MISSISSIPPI		NORTH CAROLINA		SOUTH CAROLINA		TENNESSEE		TEXAS		VIRGINIA	
	cc	sc	cc	sc	cc	sc	cc	sc	cc	sc	cc	sc	cc	sc	cc	sc	cc	sc	cc	sc	cc	sc
Size Distribution of Farm (df = 6)	3.27 (.77)[a]	3.23 (.78)	9.39 (.16)	10.49 (.11)	5.67 (.23)*	Same	7.21 (.31)	7.87 (.25)	4.46 (.48)	5.56 (.36)	6.55 (.35)	5.59 (.40)	5.70 (.46)	5.14 (.52)	2.39 (.88)	Same	8.95 (.18)	Same	7.25 (.30)	7.61 (.27)	2.85* (.42)	Same
Size Distribution of Slaveholdings (df = 12-16)	48.65 (***)	48.79 (***)	13.22 (.36)	13.84 (.32)	4.13 (.39)*		48.05 (***)	49.26 (***)	49.44 (***)	49.94 (***)	65.10 (.001)	46.35 (.001)	21.95 (.08)	21.49* (.09)	28.21 (.03)		25.88 (.03)		42.56 (***)	42.24 (***)	4.94 (.29)**	
Distribution by Age and Sex (df = 8)																						
Free Males (df = 8)	32.71 (***)	31.85 (***)	19.28 (.015)	17.48 (.025)	4.30 (.83)		50.57 (***)	49.10 (***)	5.08 (.75)	5.10 (.75)	35.98 (***)	34.04 (***)	26.77 (.001)	26.80 (.001)	33.26 (***)		35.34 (***)		12.71 (.13)	12.36 (.15)	14.61 (.07)	
Free Females (df = 8)	25.65 (.002)	25.46 (.002)	13.04 (.12)	12.97 (.12)	6.78 (.56)		33.01 (***)	32.36 (***)	5.12 (.74)	5.26 (.73)	16.66 (.04)	16.44 (.04)	10.64 (.23)	10.93 (.21)	16.93 (.03)		8.48 (.39)		11.74 (.17)	11.79 (.17)	4.15 (.84)	
Slave Males (df = 8)	11.96 (.16)	12.92 (.12)	11.21 (.19)	11.02 (.20)	12.41 (.14)		9.46 (.31)	9.35 (.31)	18.43 (.02)	15.37 (.05)	3.38 (.91)	3.00 (.93)	13.35 (.10)	12.99 (.12)	9.44 (.31)		6.36 (.61)		28.08 (***)	27.94 (***)	10.19 (.26)	
Slave Females (df = 8)	26.33 (.001)	26.55 (.001)	3.76 (.88)	3.62 (.89)	5.28 (.73)		12.46 (.14)	12.05 (.15)	26.21 (.001)	31.10 (***)	11.04 (.20)	10.28 (.25)	11.42 (.18)	10.85 (.21)	21.74 (.005)		5.54 (.70)		4.86 (.77)	4.78 (.88)	28.93 (***)	
Condition Sex (df = 3)	103.9 (***)	105.8 (***)	11.93 (.008)	9.10 (.03)	8.72 (.04)		399.5 (***)	437.41 (***)	314.05 (***)	298.74 (***)	34.98 (***)	49.58 (***)	44.55 (***)	49.06 (***)	127.70 (***)		48.83 (***)		150.17 (***)	154.92 (***)	38.32 (***)	

* (df = 3)
** (df = 4)
*** PL .0005

[a]Probability of finding a chi-square value as large as or larger than the one obtained given that the sample distribution is actually the same as that of the population.

APPENDIX VI

COUNTIES INCLUDED IN THE 1850 SAMPLE--BY REGION

	Bales	Soil Type	Farms in County	Farms in Sample
Alluvial Regions				
Alabama				
Limestone	14,809	8	649	30
Madison	20,888	8	1080	42
Arkansas				
Arkansas	3,769	5	153	6
Chicot	12,196	27	142	8
Crittenden	698	27	192	6
Deshia	2,672	27	118	6
Jefferson	4,273	27	317	14
Louisiana				
Bossier	4,181	16	333	15
Carrol	15,544	27	238	9
Concordia	18,297	27	148	4
DeSoto	2,995	16	427	15
Madison	12,771	27	218	9
Rapides	4,222	27	187	2
Tensas	21,665	27	165	omitted
Mississippi				
Adams	17,473	20	136	6
Bolivar	4,723	27	57	3
Carrol	17,989	20	986	36
Claiborne	20,795	20	310	12
Coahoma	2,430	27	161	8
Hinds	19,829	18	852	36
Issaquena	8,461	27	76	3
Jefferson	16,193	16	239	12
Madison	14,863	18	552	21
Tallahatchie	4,977	27	282	12
Tunica	717	27	41	3
Warren	18,513	20	435	15
Wilkinson	26,381	20	404	14
Yazoo	22,052	27	464	18
Total	334,372		9362	365
Other Regions of The New South				
Alabama				
Butler	4,094	23	553	15
Chambers	17,442	1	1342	21
Coosa	5,524	1	1130	21
Dallas	35,275	17	749	17
Fayette	2,920	16	1065	21
Green	25,680	17	1310	21
Lowndes	23,872	17	874	18

	Bales	Soil Type	Farms in County	Farms in Sample
Other Regions of The New South				
Marengo	32,295	17	818	15
Marion	1,552	16	573	12
Montgomery	25,326	17	962	21
Pike	8,679	23	1533	26
Texas				
Austin	3,205	19	230	3
Fort Bent	2,465	27	109	3
Grimes	2,282	16	217	3
Walker	873	16	234	6
Washington	4,003	19	306	omitted
Total	195,492		12005	223
Old South Regions				
Georgia				
Burke	19,175	23	712	9
Clay (not a county in 1850)				
Coweta	10,369	1	911	15
Dougherty (not a county in 1850)				
Hancock	11,374	1	444	9
Harris	11,935	1	873	12
Houston	19,362	23	750	12
Meriwether	12,862	1	824	12
Monroe	15,012	1	746	15
Morgan	11,541	1	336	6
Newton	6,938	1	812	12
Putnam	8,621	1	351	6
Stewart	19,165	23	990)	
Sumter	7,535	23	768) microfilm	
Thomas	7,667	23	534) missing	
Troup	14,481	1	789)	
Washington	7,445	23	632)	
South Carolina				
Abbeville	27,192	1	1814	24
Barnwell	10,138	23	1558	omitted
Chester	17,810	1	844	15
Clarendon (not a county in 1850)				
Darlington	13,005	23	857	12
Edgefield	25,880	1	2030	30
Laurens	15,842	1	1603	27
Marion	8,680	23	1374	24
Marlborough	9,501	23	621	9
Newberry	19,894	1	1045	18
Richland	11,365	4	543	6
Sumter	18,799	23	1343	18
Union	14,156	1	869	18
Total	375,744		24973	309

Source: U.S. Bureau of the Census, _Seventh Census of the United States: 1850_.

APPENDIX VII

COUNTIES INCLUDED IN THE 1860 SAMPLE--BY REGION

	Bales	Soil Type	Farms in County	Farms in Sample
Alluvial Regions				
Alabama				
Limestone	15,115	8	672	15
Madison	22,119	8	915	15
Arkansas				
Arkansas	20,178	5	453	5
Chicot	40,948	27	198	omitted
Crittenden	4,675	27	202	5
Desha	12,261	27	295	10
Jefferson	28,586	27	645	15
Louisiana				
Bossier	40,028	16	405	7
Carrol	84,165	27	513	10
Concordia	63,971	27	200	5
De Soto	16,554	16	609	10
Madison	44,870	27	215	5
Rapides	49,168	27	841	14
Tensas	141,493	27	207	5
Mississippi				
Adams	29,788	20	214	5
Bolivar	33,452	27	223	5
Carrol	42,880	20	950	20
Claiborne	33,178	20	305	5
Coahoma	13,325	27	166	omitted
Hinds	54,685	18	595	9
Issaquena	41,170	27	104	omitted
Jefferson	30,913	16	319	5
Madison	51,327	18	596	10
Tallahatchie	15,894	27	396	10
Tunica	13,025	27	112	omitted
Warren	36,338	20	396	4
Wilkinson	39,387	20	401	5
Yazoo	64,075	27	546	10
Total	1,083,568		11,693	209
Other Regions of The New South				
Alabama				
Butler	13,489	23	1027	20
Chambers	24,589	1	1288	25
Coosa	13,990	1	1631	35
Dallas	63,410	17	711	15
Fayette	5,462	16	1189	25
Green	57,858	17	793	19
Lowndes	53,664	17	1033	24
Marengo	62,428	17	778	15
Marion	4,285	16	1127	25

	Bales	Soil Type	Farms in County	Farms in Sample
Other Regions of The New South				
Montgomery	58,880	17	800	15
Pike	24,527	23	1682	30
Texas				
Austin	19,020	19	790	20
Fort Bent	13,602	27	161	5
Grimes	18,303	16	484	10
Walker	11,980	16	349	5
Washington	23,221	19	575	15
Total	468,708		14,418	303
Old South Regions				
Georgia				
Burke	23,419	23	675	15
Clay	5,292	23	222	5
Coweta	14,930	1	870	15
Dougherty	19,580	24	189	omitted
Hancock	13,332	1	374	5
Harris	14,906	1	683	15
Houston	28,852	23	487	10
Meriwether	18,159	1	685	15
Monroe	17,165	1	662	15
Morgan	8,097	1	318	5
Newton	7,983	1	777	25
Putnam	11,319	1	275	5
Stewart	25,902	23	598	10
Sumter	14,423	23	533	10
Thomas	6,582	23	299	7
Troup	17,978	1	648	10
Washington	12,421	23	697	15
South Carolina				
Abbeville	21,977	1	1233	30
Barnwell	23,490	23	1407	35
Chester	16,486	1	791	20
Clarendon	9,568	26	589	15
Darlington	16,923	23	853	20
Edgefield	27,197	1	1696	35
Laurens	15,901	1	1271	22
Marion	13,692	23	1179	25
Marlborough	13,596	23	610	15
Newberry	17,476	1	841	9
Richland	9,946	4	203	5
Sumter	18,108	23	889	19
Union	15,874	1	786	20
Total	480,574		21,340	452

Source: U.S. Bureau of the Census, Eighth Census of the United States: 1860, Agriculture.

APPENDIX VIII

SOIL TYPE CODES

Code	Description of Soil Type
1	Granite and Metamorphic gray and red lands.
2	Higland Rim of Central Basin of Tennessee.
3	Siliceous and Mountain Lands, Tenn. N. Ala., N. Ark.
4	Sand Hills Belt of middle N.C., S.C., Ga., Ala.
5	Gray Silt Prairies, Ark., La., Texas.
6	Sand Desert, S.W. Texas.
7	Red Clay Lands, Middle Tennessee.
8	Valley Lands of E. Tenn., Ga., Ala. with narrow cherty ridges.
9	Red Loam and Clay Prairie Country, Ark., Mo., Texas.
10	Pontotoc Ridge Red Lands (Cretaceous), Miss.
11	Red Hills (Tertiary), S.C.
12	Central Basin Limestone Lands, Tenn.
13	Red Loam Region, Ark.
14	Flatwoods, Ala., Miss., Tenn.
15	Gypsum Prairies and arid Plains of Llano Estacado of Texas.
16	Oak, Hickory and short-leaf Pine Uplands, Miss., Tenn., Ark., Tex.
17	Black & stiff Calcareous (Cretaceous) Prairies, Ala., Miss., Ark., Tex.
18	Calcareous Prairielands (Tertiary) Ala., Miss., La., Tex.
19	Black calcareous and coast Prairies, Texas, La.
20	Brown Loam Bluff and Table Land, Tenn., Miss.
21	Brown Loam Prairies of La.

22 Sandy Prairies of Indian Territory.

23 Central Belt Oak, Hickory and long-leaf Pine Hills, N.C. to La.

24 Long-leaf Pine Hills, Ala., Fla., Ga., N.C., Tex. to Miss., and level Wine-grass and Pine Barrens.

25 Gray, gravelly Ridges and Magnesium Limestone Lands, Ark., Tenn., N. W. Ga.

26 Long-leaf Pine Flats & Savannahs near the coast from Va. to Tex.

27 Alluvial Lands and large Upland Swamps.

28 Marshes, Swamps and Live Oak Lands of the coast.

Source: U.S. Bureau of the Census, Tenth Census of the United States, 1880, Vol. V, "Cotton Production in the United States," General Agricultural Map for all the States, opposite p. 3.

BIBLIOGRAPHY

BIBLIOGRAPHY

Public Documents

U.S. Bureau of the Census. *Eighth Census of the United States: 1860, Agriculture.*

U.S. Bureau of the Census. *Eighth Census of the United States: 1860, Population.*

U.S. Bureau of the Census. *Historical Statistics of the United States, Colonial Times to 1957.* Washington: U.S. Government Printing Office, 1960.

U.S. Bureau of the Census. *Seventh Census of the United States: 1850.*

U.S. Bureau of the Census. *Seventh Census of the United States: 1850, Compendium.*

U.S. Bureau of the Census. *Tenth Census of the United States: 1880, Cotton Production in the United States*, Vol. V.

Books

Adams, James Truslow. *Atlas of American History.* New York: Charles Scribner's Sons, 1943.

Baldwin, J. G. *The Flush Times of Alabama and Mississippi.* New York: Sagamore Press, 1957.

Cairnes, J. E. *The Slave Power: Its Character, Career and Probable Designs.* 2nd ed. London: Macmillan and Co., 1863.

Cochran, William G. *Sampling Techniques.* 2nd ed. revised. New York: John Wiley & Sons, Inc., 1964.

Cole, Arthur Harrison. *Wholesale Commodity Prices in the U. S. 1700-1861.* Cambridge: Harvard University Press, 1938.

Conrad, Alfred H. and Meyer, John R. *The Economics of Slavery and Other Studies in Econometric History.* Chicago: Aldine, 1964.

Croxton, Frederick E. and Cowden, Dudley J. *Practical Business Statistics.* 3rd ed. revised. Englewood Cliffs: Prentice-Hall, Inc., 1960.

Dixon, Wilfrid J. and Massey, Frank J. Jr. *Introduction to Statistical Analysis*. 2nd ed. revised. New York: McGraw-Hill Book Company, Inc., 1957.

Genovese, Eugene D. The Political Economy of Slavery: *Studies in the Economy and Society of the Slave South*. New York: Pantheon Books, 1965.

Gray, Lewis C. *History of Agriculture in the Southern United States to 1860*. 2 vols. New York: Peter Smith, 1941.

Kemble, Francis Anne. *Journal of a Residence on a Georgia Plantation in 1838-39*. New York: Alfred A. Knopf, 1961.

Levy, Lester S. and Sampson, Roy J. *American Economic Development*. Boston: Allyn and Bacon, 1963.

Mitchell, Broadus. *William Gregg, Factory Master of the Old South*. Chapel Hill: The University of North Carolina Press, 1928.

North, Douglass C. *Growth and Welfare in the American Past: A New Economic History*. Englewood Cliffs: Prentice-Hall, Inc., 1966.

Olmsted, Frederick Law. *The Cotton Kingdom*. New York: Alfred A. Knopf, 1953.

Owsley, Frank L. *Plain Folk of the Old South*. Chicago: Quadrangle Books, 1965.

Paxson, Frederic L. *The History of the American Frontier*. Boston: Houghton Mifflin Company, 1924.

Phillips, Ulrich B. *American Negro Slavery*. New York: D. Appleton and Company, 1918.

_______. *A Documentary History of American Industrial Society*. Vol. II: *Plantation and Frontier*. Cleveland: The Arthur H. Clark Company, 1910.

_______. *Life and Labor in the Old South*. Boston: Little, Brown and Company, 1963.

Robertson, Ross M. *History of the American Economy*. 2nd ed. New York: Harcourt, Brace & World, Inc., 1964.

Shannon, Fred A. *The Economic History of the United States*. Vol. V: *The Farmer's Last Frontier, Agriculture, 1860-1879*. New York: Rinehart & Company, Inc., 1959.

Stampp, Kenneth M. *The Peculiar Institution: Slavery in the Ante-Bellum South*. New York: Alfred A. Knopf, 1956.

Turner, Frederick Jackson. Frontier and Section. Englewood Cliffs: Prentice-Hall, Inc., 1961.

______. The Frontier in American History. New York: Henry Holt and Company, 1940.

Woodman, Harold D. Slavery and the Southern Economy. New York: Harcourt, Brace & World, 1966.

Wright, Carroll D. History and Growth of the United States Census. Washington: Government Printing Office, 1900.

Articles

Conrad, Alfred H. and Meyer, John R. "The Economics of Slavery in the Antebellum South," Journal of Political Economy, LXVI (April, 1958), 95-122.

Danhof, Clarence H. "Economic Validity of the Safety-Valve Doctrine," The Journal of Economic History, I (March, 1941), 96-106.

Linden, Fabian. "Economic Democracy in the Slave South: An Appraisal of Some Recent Views," Journal of Negro History, XXXI (April, 1946), 140-89.

Russel, Robert R. "The Effects of Slavery upon Nonslaveholders in the Ante-Bellum South," Agricultural History, XV (April, 1941), 112-26.

Saraydar, Edward. "A Note on the Profitability of Ante Bellum Slavery," Southern Economic Journal, XXX (April, 1964), 325-32.

Sutch, Richard. "The Profitability of Ante Bellum Slavery--Revisited," Southern Economic Journal, XXXI (April, 1965), 365-83.

Unpublished Material

U.S. Bureau of the Census. Eighth Census of the United States: 1860, Manuscript Census Schedules of Agriculture.

U.S. Bureau of the Census. Eighth Census of the United States: 1860, Manuscript Census Schedules of Free Inhabitants.

U.S. Bureau of the Census. Eighth Census of the United States: 1860, Manuscript Census Schedules of Slave Inhabitants.

U.S. Bureau of the Census. Seventh Census of the United States: 1850, Manuscript Census Schedules of Agriculture.

U.S. Bureau of the Census. *Seventh Census of the United States: 1850, Manuscript Census Schedules of Free Inhabitants.*

U.S. Bureau of the Census. *Seventh Census of the United States: 1850, Manuscript Census Schedules of Slave Inhabitants.*

Dissertations in American Economic History

An Arno Press Collection

Adams, Donald R., Jr. **Wage Rates in Philadelphia, 1790-1830.** (Doctoral Dissertation, University of Pennsylvania, 1967). 1975

Aldrich, Terry Mark. **Rates of Return on Investment in Technical Education in the Ante-Bellum American Economy.** (Doctoral Dissertation, The University of Texas at Austin, 1969). 1975

Anderson, Terry Lee. **The Economic Growth of Seventeenth Century New England:** A Measurement of Regional Income. (Doctoral Dissertation, University of Washington, 1972). 1975

Bean, Richard Nelson. **The British Trans-Atlantic Slave Trade, 1650-1775.** (Doctoral Dissertation, University of Washington, 1971). 1975

Brock, Leslie V. **The Currency of the American Colonies, 1700-1764:** A Study in Colonial Finance and Imperial Relations. (Doctoral Dissertation, University of Michigan, 1941). 1975

Ellsworth, Lucius F. **Craft to National Industry in the Nineteenth Century:** A Case Study of the Transformation of the New York State Tanning Industry. (Doctoral Dissertation, University of Delaware, 1971). 1975

Fleisig, Heywood W. **Long Term Capital Flows and the Great Depression:** The Role of the United States, 1927-1933. (Doctoral Dissertation, Yale University, 1969). 1975

Foust, James D. **The Yeoman Farmer and Westward Expansion of U. S. Cotton Production.** (Doctoral Dissertation, University of North Carolina at Chapel Hill, 1968). 1975

Golden, James Reed. **Investment Behavior By United States Railroads, 1870-1914.** (Doctoral Thesis, Harvard University, 1971). 1975

Hill, Peter Jensen. **The Economic Impact of Immigration into the United States.** (Doctoral Dissertation, The University of Chicago, 1970). 1975

Klingaman, David C. **Colonial Virginia's Coastwise and Grain Trade.** (Doctoral Dissertation, University of Virginia, 1967). 1975

Lang, Edith Mae. **The Effects of Net Interregional Migration on Agricultural Income Growth:** The United States, 1850-1860. (Doctoral Thesis, The University of Rochester, 1971). 1975

Lindley, Lester G. **The Constitution Faces Technology:** The Relationship of the National Government to the Telegraph, 1866-1884. (Doctoral Thesis, Rice University, 1971). 1975

Lorant, John H[erman]. **The Role of Capital-Improving Innovations in American Manufacturing During the 1920's.** (Doctoral Thesis, Columbia University, 1966). 1975

Mishkin, David Joel. **The American Colonial Wine Industry:** An Economic Interpretation, Volumes I and II. (Doctoral Thesis, University of Illinois, 1966). 1975

Oates, Mary J. **The Role of the Cotton Textile Industry in the Economic Development of the American Southeast:** 1900-1940. (Doctoral Dissertation, Yale University, 1969). 1975

Passell, Peter. **Essays in the Economics of Nineteenth Century American Land Policy.** (Doctoral Dissertation, Yale University, 1970). 1975

Pope, Clayne L. **The Impact of the Ante-Bellum Tariff on Income Distribution.** (Doctoral Dissertation, The University of Chicago, 1972). 1975

Poulson, Barry Warren. **Value Added in Manufacturing, Mining, and Agriculture in the American Economy From 1809 To 1839.** (Doctoral Dissertation, The Ohio State University, 1965). 1975

Rockoff, Hugh. **The Free Banking Era: A Re-Examination.** (Doctoral Dissertation, The University of Chicago, 1972). 1975

Schumacher, Max George. **The Northern Farmer and His Markets During the Late Colonial Period.** (Doctoral Dissertation, University of California at Berkeley, 1948). 1975

Seagrave, Charles Edwin. **The Southern Negro Agricultural Worker:** 1850-1870. (Doctoral Dissertation, Stanford University, 1971). 1975

Solmon, Lewis C. **Capital Formation by Expenditures on Formal Education, 1880 and 1890.** (Doctoral Dissertation, The University of Chicago, 1968). 1975

Swan, Dale Evans. **The Structure and Profitability of the Antebellum Rice Industry:** 1859. (Doctoral Dissertation, University of North Carolina at Chapel Hill, 1972). 1975

Sylla, Richard Eugene. **The American Capital Market, 1846-1914:** A Study of the Effects of Public Policy on Economic Development. (Doctoral Thesis, Harvard University, 1968) 1975

Uselding, Paul John. **Studies in the Technological Development of the American Economy During the First Half of the Nineteenth Century.** (Doctoral Dissertation, Northwestern University, 1970) 1975

Walsh, William D[avid]. **The Diffusion of Technological Change in the Pennsylvania Pig Iron Industry, 1850-1870.** (Doctoral Dissertation, Yale University, 1967). 1975

Weiss, Thomas Joseph. **The Service Sector in the United States, 1839 Through 1899.** (Doctoral Thesis, University of North Carolina at Chapel Hill, 1967). 1975

Zevin, Robert Brooke. **The Growth of Manufacturing in Early Nineteenth Century New England.** 1975